ECOLOGY
AND
ENVIRONMENT

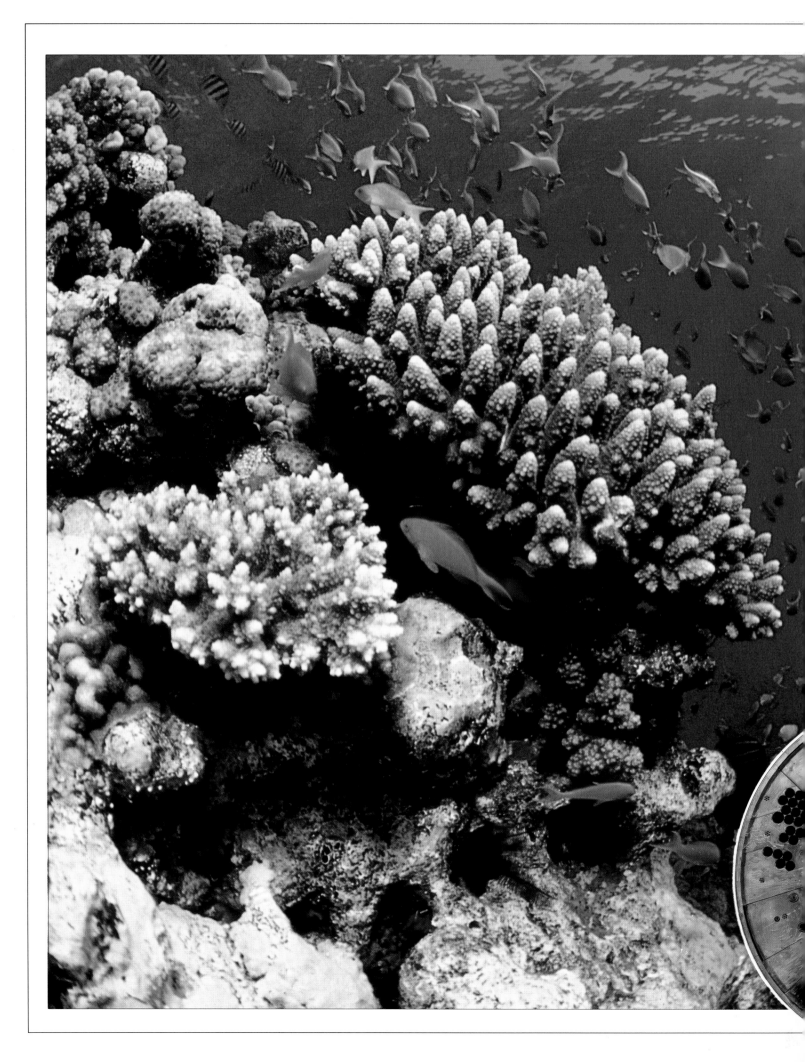

ECOLOGY AND ENVIRONMENT
The Cycles of Life

SALLY MORGAN

OXFORD UNIVERSITY PRESS

New York 1995

CONTENTS

Project editor   Peter Furtado
Editors   Lauren Bourque, John Clark
Editorial assistant   Marian Dreier

Art editor   Ayala Kingsley
Visualization and artwork   Ted McCausland/ Siena Artworks
Senior designer   Martin Anderson
Designer   Roger Hutchins

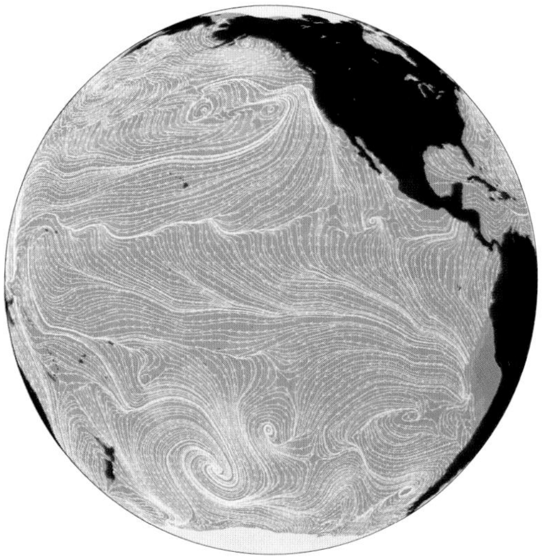

Picture manager   Jo Rapley
Picture research   Alison Floyd
Production   Clive Sparling

Planned and produced by
Andromeda Oxford Ltd
9-15 The Vineyard
Abingdon
Oxfordshire OX14 3PX

© copyright Andromeda Oxford Ltd 1995

Text pages 16–47
© copyright Helicon Ltd,
adapted by Andromeda Oxford Ltd

Published in the United States of America by
Oxford University Press, Inc.,
198 Madison Avenue
New York, NY 10016

Oxford is a resitered trademark of Oxford University Press

Library of Congress Cataloging-in-Publication Data

Morgan, Sally
    Ecology and environment : the cycles of life   / by Sally Morgan
    160pp. 29 x 23cm - (New encyclopedia of science)
    Includes bibliographical references (p.152) and index
    ISBN 0–19–521140–5
    1. Ecology   2. Environmental sciences   I. Title II. Series
QH541.M565   1995
574.5–dc20   95–17140

Printed in Spain by Graficromo SA, Córdoba

INTRODUCTION

THE EARTH has been home to human beings for about 40,000 years – an insignificant amount of time compared with the more than 3 billion years in which life has flourished. In a mere 150 years or so, our newcomer species has done as much to alter the planet as the meteor that is thought to have struck the Earth 65 million years ago, causing mass extinctions of both plants and animals, and profound climate change. Then the Earth itself recovered and flourished again, but it is now at a critical point, beyond which there may be no recovery if exploitation by humans continues on the present scale.

There are several reasons for the unprecedented impact of our species on our environment. The first is the sheer number of people that now occupy the planet. The human population was steady for most of our history, but since AD 1800 it has grown explosively. With 70 percent of the planet covered by water, and another 20 percent covered by ice, desert or steep mountains, there is very little room to support these burgeoning numbers, many of whom are desperately poor. This is the chief pressure that leads to the destruction of valuable natural systems such as forests and wetlands, as people clear the vegetation to provide living space and farm or pastureland. As these natural systems disappear, so do the wild plants and animals that occupy them.

Another key aspect is the over-use of resources. Our food, shelter, clothing and industrial materials are taken from the Earth's air, water, minerals, plants and animals. Many of these resources are recycled by natural processes. However, some (such as water and oil) are being used faster than they can be replenished, while others (such as metals) are consumed and thrown away. This is a particular problem in industrialized countries, which account for less than 25 percent of the world's nations, but consume by far the greatest share of resources. All species could survive the exhaustion of some resources, such as fossil fuels; fewer would survive if resources such as topsoil were destroyed; fewer still would survive the loss of the planet's oxygen supply.

Another problem – particularly in industrialized countries, but increasingly in developing countries as well – is pollution. Chemical products of industry, along with sewage, have been dumped onto the land and into the air and water in such large quantities, or at such highly toxic levels, that the Earth's natural filtering and diluting systems are unable to break them down and flush them out. Chlorofluorocarbons (CFCs) are letting in higher levels of harmful ultraviolet radiation from the Sun, and excess carbon dioxide gas is trapping more heat close to Earth, slowly raising the planet's temperature. The disruption of the global climate that is likely to result will have far-ranging effects on all species, although the difficulty of making accurate climate models makes it difficult to predict how much the temperature may rise, and how severe the climate disruption may be.

Ecology seeks solutions to these urgent problems by understanding their full context: exactly how human demands on the Earth affect its natural systems (ecosystems), and how these systems inter-relate with each other and with all the species they support. As these issues become better understood, the way to balance the continued progress of human civilization with the survival of the Earth on which it depends will become more clear.

T HIS BOOK aims to make all this information available to the whole family, from students studying for examinations and projects to adults wanting to bring their scientific knowledge up to date. To achieve this, the book is organized in such a way as to provide readers with a quick answer to a specific query, or allow them to follow a more detailed account of a particular topic.

At the heart of the book is a 96-page thematic section, made up of 48 major narrative topics, each one richly illustrated to tell the story of a central theme of the book. The strong graphic presentation and the style of writing are designed to make this section the ideal point of departure for the less well-informed reader. Sets of Keywords highlighted on each topic spread point the reader to the second major section, a 32-page alphabetic mini-encyclopedia of ecology, which contains some 400 entries. This section, too, leads the reader back to the thematic topics.

No part of modern science can be separated cleanly from other fields. Ecology and environmental studies merge into biology on the one hand, geology on the other and genetics in other respects. It has also illuminated aspects of modern economics and other social sciences. The Knowledge Map, immediately following this Introduction, maps out the entire field of modern science, shows how each area of science interacts with another, and defines the major fields. This is followed by a Timechart, which traces the development of the subject through great discoveries.

Finally, to ensure that the volume is of genuine value for reference as well as for recreational browsing, the Factfile provides a wealth of hard data, maps, tables, lists and statistics.

KNOWLEDGE MAP
Key Fields of Modern Science

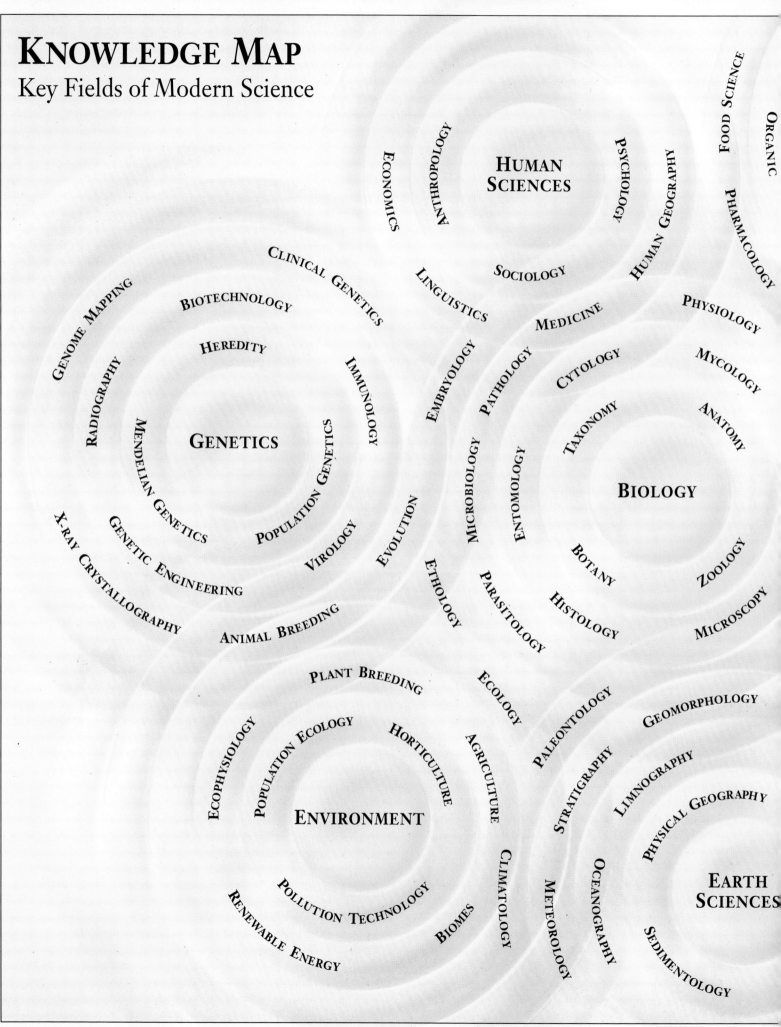

HUMAN SCIENCES

ECONOMICS · ANTHROPOLOGY · PSYCHOLOGY · HUMAN GEOGRAPHY · FOOD SCIENCE · ORGANIC · PHARMACOLOGY · LINGUISTICS · SOCIOLOGY · MEDICINE · PHYSIOLOGY

GENETICS

GENOME MAPPING · CLINICAL GENETICS · BIOTECHNOLOGY · HEREDITY · RADIOGRAPHY · IMMUNOLOGY · MENDELIAN GENETICS · POPULATION GENETICS · GENETIC ENGINEERING · VIROLOGY · X-RAY CRYSTALLOGRAPHY · ANIMAL BREEDING · EMBRYOLOGY · EVOLUTION

BIOLOGY

MYCOLOGY · ANATOMY · CYTOLOGY · TAXONOMY · PATHOLOGY · ENTOMOLOGY · MICROBIOLOGY · BOTANY · ZOOLOGY · HISTOLOGY · MICROSCOPY · PARASITOLOGY · ETHOLOGY · ECOLOGY

ENVIRONMENT

PLANT BREEDING · ECOPHYSIOLOGY · POPULATION ECOLOGY · HORTICULTURE · AGRICULTURE · POLLUTION TECHNOLOGY · RENEWABLE ENERGY · BIOMES · CLIMATOLOGY · PALEONTOLOGY

EARTH SCIENCES

GEOMORPHOLOGY · STRATIGRAPHY · LIMNOGRAPHY · PHYSICAL GEOGRAPHY · OCEANOGRAPHY · METEOROLOGY · SEDIMENTOLOGY

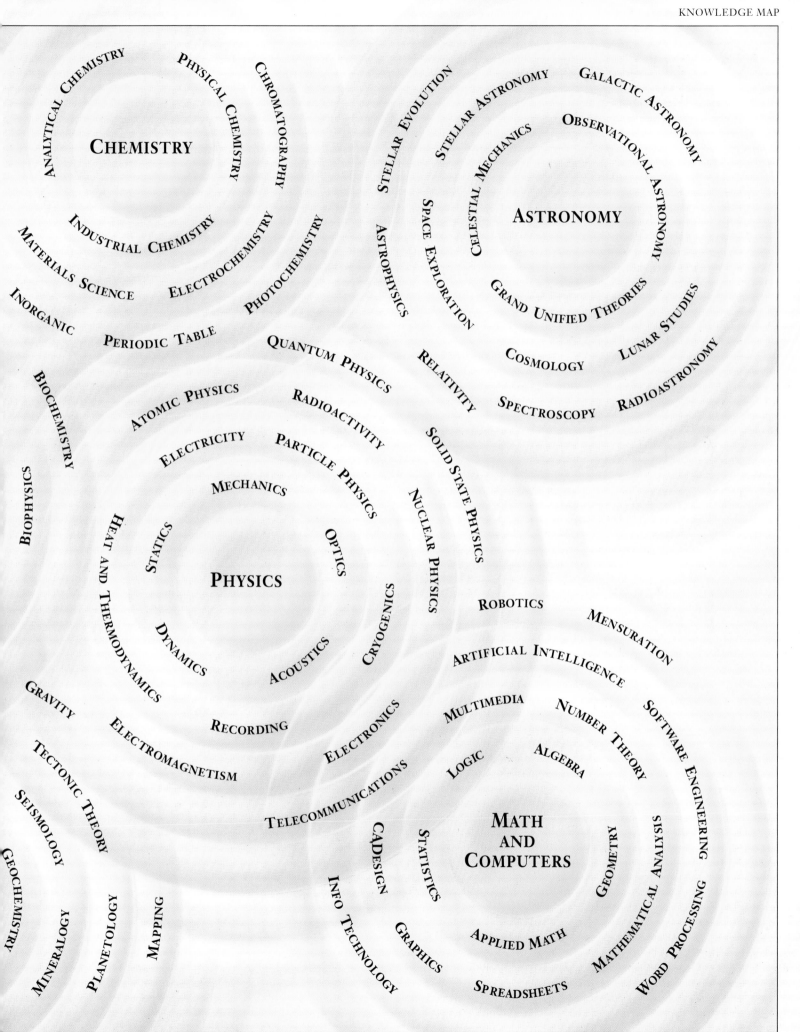

KNOWLEDGE MAP
Modern Ecology

AGROECOLOGY

A branch of agriculture that incorporates ideas about a more environmentally sensitive approach, focusing on sustainability of production rather than yield. The crop field is considered to be an ecosystem in which processes such as nutrient recycling, predator-prey relationships and crop-weed competition occur.

HORTICULTURE

The science of growing plants. Commercial horticulture developed with industrial towns in the 19th century. Now it makes use of greenhouses with optimized light levels, temperature, water and nutrient levels.

POPULATION ECOLOGY

The study of populations – the numbers of a group of living things. Populations change over a period of time as a result of births, deaths and migration. The size of a population is limited by a number of factors. These include competition for food, living space, oxygen and water; the presence of predators and disease-causing agents.

ECOLOGY

The study of the interactions that determine the distribution of all living things in their environment. Ecology has two branches. The study of the relationship between individual organisms or species and their surroundings is called autecology; by contrast, synecology is concerned with the whole community of plants and animals and their environment.

ANIMAL BREEDING

The study of the techniques that have enabled farmers to improve domesticated animals' yields of meat, eggs and milk. Modern livestock tend to be larger than their ancestors, maturing earlier and yielding more product. Gene manipulation may enable animals to produce new products, such as milk with medicinal drugs.

AGRICULTURE

The cultivation of crops and the rearing of domesticated animals. The origins of agriculture go back 9000 years. Since that time, people have manipulated their own environment to suit the needs of animals and crops. Today there are two main types of agriculture worldwide: subsistence, in which farmers plant crops for their own consumption; and cash, in which farmers grow crops or rear animals specifically in order to sell them to other people.

PLANT BREEDING

The study of the modification of plants by artificial selection. Plants have been modified by people or centuries, and the practice is increasingly sophisticated. Modern wheats, for example, yield more grain than their ancestral species. As understanding of genetics, molecular biology and tissue culture improves, it has become possible to apply these to plant breeding. New strains of wheat may be given resistance to disease.

POPULATION GENETICS

A branch of genetics which involves the study of the genes carried by a population of organisms. There may be millions if not billions of genes in any population; the sum of these is called the gene pool. The frequency with which particular genes occur within the gene pool varies as environmental conditions change.

EVOLUTION

The study of the origin of species through gradual changes in ancestral groups. The most widely accepted theory today is Darwin's theory of evolution, in which the process of natural selection acts on randomly-occurring variations.

BIOREMEDIATION

The study of the use of organisms to clean up pollution – for example, using certain bacteria to clean up oil spills and to digest toxic dumps. Under optimal conditions, biodegradation occurs rapidly.

ECOTOXICOLOGY

The study of the harmful effects of chemicals in ecosystems. This includes the way in which the chemicals enter the ecosystem, how they are distributed through the food webs and then metabolized. Population changes are monitored with any development of resistance – for example, pests becoming resistant to pesticides.

POLLUTION TECHNOLOGY

The application of modern technology to pollution control and reduction. Techniques include sensors to determine the source of air and water pollution, and satellites to track algal blooms and oil spills at sea.

WASTE MANAGEMENT

The study of the ways in which waste produced by modern societies can be disposed of without threatening the environment. Many of the waste materials can be recycled; waste reduction is an important aspect of waste management.

CLIMATOLOGY

The study of climate, the regular pattern of weather experienced in a particular place over a period of time. The primary factors affecting climatic variation are latitude, wind movements, ocean currents, the temperature difference between land and sea, and topography. Human activities have an increasing effect.

OCEANOGRAPHY

The study of the oceans, their origin and structure, and of the marine community of plants and animals. Oceanography also involves the study of the composition of seawater and its movement in currents, waves and tides. This relatively new science is becoming increasingly important because it has been recognized that oceans have a great influence on the biosphere, particularly on the climate.

METEOROLOGY

The study of the atmosphere to forecast the weather. Data from weather stations and satellites – atmospheric pressure, air temperature, humidity, rainfall, wind speed – are collated by computer to produce a forecast.

RENEWABLE ENERGY

The study of sources of energy that replenish themselves. Most sources of renewable energy are based on solar energy, either directly or via wind power, hydroelectric power, wave power and biomass (the products of photosynthesis). Geothermal power harnesses heat from deep in the ground or from tidal power.

CONSERVATION

The study and practice of the ways in which species can be preserved with their habitats. There are three main approaches to conservation: by legislation against the sale of their produce; by setting aside habitats as managed national parks or nature reserves; and by taking threatened species into captive-breeding programs.

BIOTECHNOLOGY

The applied science of biology. Most commonly it refers to the production of living organisms for industrial, medical or other uses, as in genetically-engineered plants or bacteria. The term can also be used to refer to techniques for the provision of sustainable long-term solutions to environmental problems based on improved knowledge of ecosystems – for example, renewable energy and reforestation.

TIMECHART

MANY of the first formal observations of the natural world, such as meteorology, were made in ancient Greece, although some Chinese records of meteorological events date back to 1000 BC The Greeks correctly related climate to latitudinal bands and day-to-day weather to different winds and air masses. Aristotle (384–322 BC) wrote on meteorology, and his pupil Theophrastus (380–285 BC) produced a collection of forecasting rules and weather lore, as well as two seminal works on botany. Many Greek authors, including Herodotus (c.484–c.425 BC), Xenophon (c.430–c.355 BC), Strabo (c.63 BC–c.AD 21) and Aristotle, remarked on human impacts on the environment, including deforestation for agriculture and soil erosion.

The next major phase of advancement in human study of the natural world occurred during the European colonial expansion after AD 1500. This took agriculture to huge areas of previously uncultivated land in North and South America, Australia and New Zealand. Natural scientists such as the German Alexander von Humboldt (1769–1859), and Britons Sir Joseph Banks

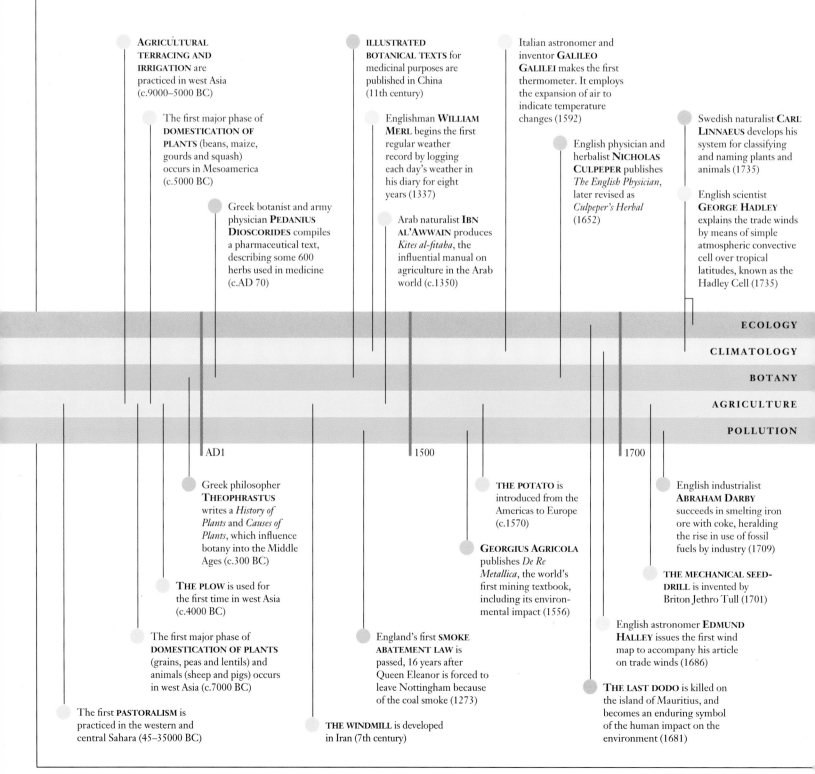

AGRICULTURAL TERRACING AND IRRIGATION are practiced in west Asia (c.9000–5000 BC)

The first major phase of **DOMESTICATION OF PLANTS** (beans, maize, gourds and squash) occurs in Mesoamerica (c.5000 BC)

Greek botanist and army physician **PEDANIUS DIOSCORIDES** compiles a pharmaceutical text, describing some 600 herbs used in medicine (c.AD 70)

ILLUSTRATED BOTANICAL TEXTS for medicinal purposes are published in China (11th century)

Englishman **WILLIAM MERL** begins the first regular weather record by logging each day's weather in his diary for eight years (1337)

Arab naturalist **IBN AL'AWWAIN** produces *Kites al-fitaha*, the influential manual on agriculture in the Arab world (c.1350)

Italian astronomer and inventor **GALILEO GALILEI** makes the first thermometer. It employs the expansion of air to indicate temperature changes (1592)

English physician and herbalist **NICHOLAS CULPEPER** publishes *The English Physician*, later revised as *Culpeper's Herbal* (1652)

Swedish naturalist **CARL LINNAEUS** develops his system for classifying and naming plants and animals (1735)

English scientist **GEORGE HADLEY** explains the trade winds by means of simple atmospheric convective cell over tropical latitudes, known as the Hadley Cell (1735)

ECOLOGY

CLIMATOLOGY

BOTANY

AGRICULTURE

POLLUTION

AD1

1500

1700

Greek philosopher **THEOPHRASTUS** writes a *History of Plants* and *Causes of Plants*, which influence botany into the Middle Ages (c.300 BC)

THE PLOW is used for the first time in west Asia (c.4000 BC)

The first major phase of **DOMESTICATION OF PLANTS** (grains, peas and lentils) and animals (sheep and pigs) occurs in west Asia (c.7000 BC)

The first **PASTORALISM** is practiced in the western and central Sahara (45–35000 BC)

England's first **SMOKE ABATEMENT LAW** is passed, 16 years after Queen Eleanor is forced to leave Nottingham because of the coal smoke (1273)

THE WINDMILL is developed in Iran (7th century)

THE POTATO is introduced from the Americas to Europe (c.1570)

GEORGIUS AGRICOLA publishes *De Re Metallica*, the world's first mining textbook, including its environmental impact (1556)

English industrialist **ABRAHAM DARBY** succeeds in smelting iron ore with coke, heralding the rise in use of fossil fuels by industry (1709)

THE MECHANICAL SEED-DRILL is invented by Briton Jethro Tull (1701)

English astronomer **EDMUND HALLEY** issues the first wind map to accompany his article on trade winds (1686)

THE LAST DODO is killed on the island of Mauritius, and becomes an enduring symbol of the human impact on the environment (1681)

(1743–1820), Alfred Russell Wallace (1823–1913) and Charles Darwin (1809–1882) traveled the world to observe and collect specimens and develop theories. Darwin's work on evolution had been helped by methods of biological classification developed by Swede Carl Linnaeus (1707–1778) and Frenchman Antoine Laurent de Jussieu (1748–1836). Von Humboldt paved the way for modern climatology; some of his views on relations between people and the environment anticipated today's conservationist ideas and were strongly influenced by Hindu philosophy.

Interest in understanding atmospheric processes was related to the colonial expansion of travel and trade and, eventually, to production from agricultural plantations. Botany, however, had advanced little from the Greeks until plant anatomy developed in the 17th century following the invention of the microscope at the end of the 16th century. Even in this period, interest in botany remained focused on the medicinal properties of plants.

With industrial growth and innovation in Western Europe from the mid-18th century came new problems of pollution. New

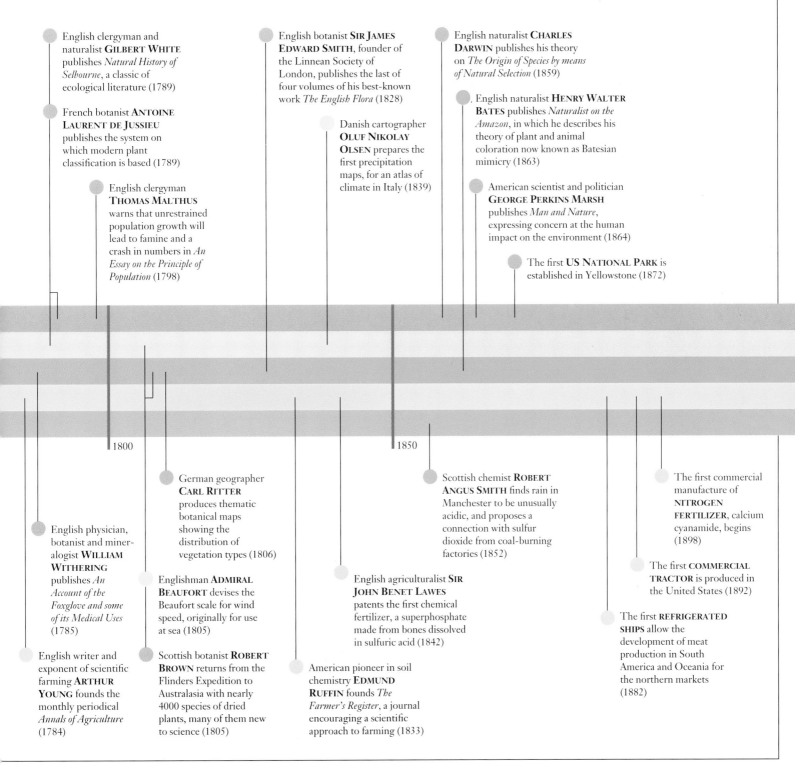

English clergyman and naturalist **GILBERT WHITE** publishes *Natural History of Selbourne*, a classic of ecological literature (1789)

French botanist **ANTOINE LAURENT DE JUSSIEU** publishes the system on which modern plant classification is based (1789)

English clergyman **THOMAS MALTHUS** warns that unrestrained population growth will lead to famine and a crash in numbers in *An Essay on the Principle of Population* (1798)

English botanist **SIR JAMES EDWARD SMITH**, founder of the Linnean Society of London, publishes the last of four volumes of his best-known work *The English Flora* (1828)

Danish cartographer **OLUF NIKOLAY OLSEN** prepares the first precipitation maps, for an atlas of climate in Italy (1839)

English naturalist **CHARLES DARWIN** publishes his theory on *The Origin of Species by means of Natural Selection* (1859)

English naturalist **HENRY WALTER BATES** publishes *Naturalist on the Amazon*, in which he describes his theory of plant and animal coloration now known as Batesian mimicry (1863)

American scientist and politician **GEORGE PERKINS MARSH** publishes *Man and Nature*, expressing concern at the human impact on the environment (1864)

The first **US NATIONAL PARK** is established in Yellowstone (1872)

1800

1850

English physician, botanist and miner-alogist **WILLIAM WITHERING** publishes *An Account of the Foxglove and some of its Medical Uses* (1785)

English writer and exponent of scientific farming **ARTHUR YOUNG** founds the monthly periodical *Annals of Agriculture* (1784)

German geographer **CARL RITTER** produces thematic botanical maps showing the distribution of vegetation types (1806)

Englishman **ADMIRAL BEAUFORT** devises the Beaufort scale for wind speed, originally for use at sea (1805)

Scottish botanist **ROBERT BROWN** returns from the Flinders Expedition to Australasia with nearly 4000 species of dried plants, many of them new to science (1805)

English agriculturalist **SIR JOHN BENET LAWES** patents the first chemical fertilizer, a superphosphate made from bones dissolved in sulfuric acid (1842)

American pioneer in soil chemistry **EDMUND RUFFIN** founds *The Farmer's Register*, a journal encouraging a scientific approach to farming (1833)

Scottish chemist **ROBERT ANGUS SMITH** finds rain in Manchester to be unusually acidic, and proposes a connection with sulfur dioxide from coal-burning factories (1852)

The first commercial manufacture of **NITROGEN FERTILIZER**, calcium cyanamide, begins (1898)

The first **COMMERCIAL TRACTOR** is produced in the United States (1892)

The first **REFRIGERATED SHIPS** allow the development of meat production in South America and Oceania for the northern markets (1882)

machinery brought about advances in agricultural technology. The work of Austrian monk Gregor Mendel (1822–1884) on plant genetics led to progress in breeding new crop varieties, although his findings did not become generally known until the early 20th century. Agricultural and industrial revolutions, together with medical advances, brought about rapid population growth and a rising rate of environmental change.

The word "ecology" was first coined in the late 19th century by German Ernst Haeckel (1834–1919) although it represented a type of thinking of considerable antiquity. The pioneering work of American George Perkins Marsh (1801–1882) foreshadowed the late 20th-century conservation movement. Connections between climate and vegetation were made by American Frederic Clements (1874–1945) in the field of ecology and German Wladimir Koeppen in climatology.

As world population grew, agriculture initially kept pace by expanding the area under cultivation. After the 1950s, intensive methods of cultivation became widely used. Chemical fertilizers,

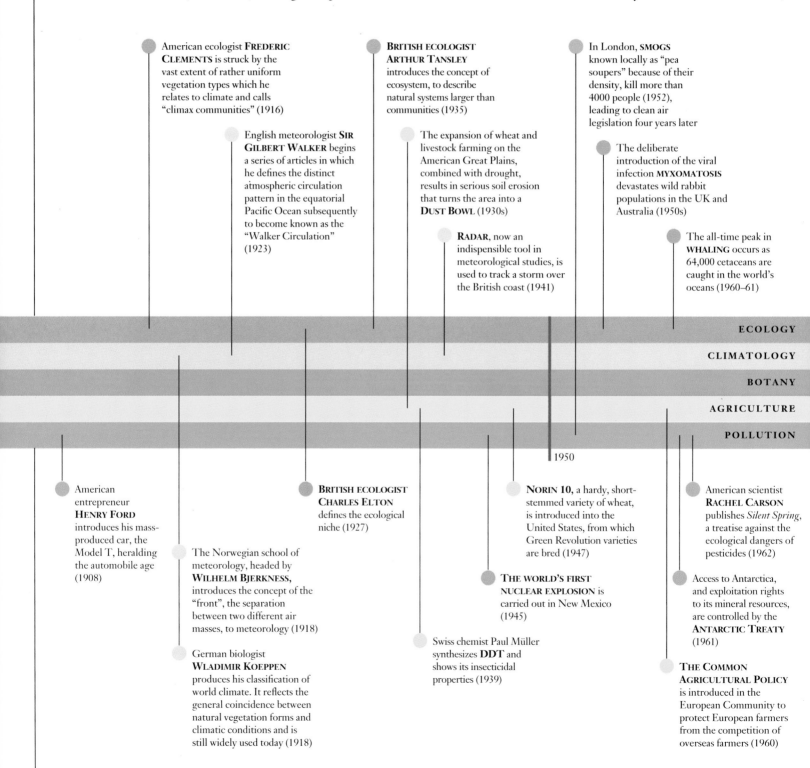

American ecologist **FREDERIC CLEMENTS** is struck by the vast extent of rather uniform vegetation types which he relates to climate and calls "climax communities" (1916)

English meteorologist **SIR GILBERT WALKER** begins a series of articles in which he defines the distinct atmospheric circulation pattern in the equatorial Pacific Ocean subsequently to become known as the "Walker Circulation" (1923)

BRITISH ECOLOGIST ARTHUR TANSLEY introduces the concept of ecosystem, to describe natural systems larger than communities (1935)

The expansion of wheat and livestock farming on the American Great Plains, combined with drought, results in serious soil erosion that turns the area into a **DUST BOWL** (1930s)

RADAR, now an indispensible tool in meteorological studies, is used to track a storm over the British coast (1941)

In London, **SMOGS** known locally as "pea soupers" because of their density, kill more than 4000 people (1952), leading to clean air legislation four years later

The deliberate introduction of the viral infection **MYXOMATOSIS** devastates wild rabbit populations in the UK and Australia (1950s)

The all-time peak in **WHALING** occurs as 64,000 cetaceans are caught in the world's oceans (1960–61)

ECOLOGY

CLIMATOLOGY

BOTANY

AGRICULTURE

POLLUTION

1950

American entrepreneur **HENRY FORD** introduces his mass-produced car, the Model T, heralding the automobile age (1908)

The Norwegian school of meteorology, headed by **WILHELM BJERKNESS,** introduces the concept of the "front", the separation between two different air masses, to meteorology (1918)

German biologist **WLADIMIR KOEPPEN** produces his classification of world climate. It reflects the general coincidence between natural vegetation forms and climatic conditions and is still widely used today (1918)

BRITISH ECOLOGIST CHARLES ELTON defines the ecological niche (1927)

Swiss chemist Paul Müller synthesizes **DDT** and shows its insecticidal properties (1939)

NORIN 10, a hardy, short-stemmed variety of wheat, is introduced into the United States, from which Green Revolution varieties are bred (1947)

THE WORLD'S FIRST NUCLEAR EXPLOSION is carried out in New Mexico (1945)

American scientist **RACHEL CARSON** publishes *Silent Spring*, a treatise against the ecological dangers of pesticides (1962)

Access to Antarctica, and exploitation rights to its mineral resources, are controlled by the **ANTARCTIC TREATY** (1961)

THE COMMON AGRICULTURAL POLICY is introduced in the European Community to protect European farmers from the competition of overseas farmers (1960)

pesticides and herbicides, and development of new varieties of crop plants enabled an explosion in crop yields, often referred to as the "Green Revolution", in the tropical world. Agrochemicals, first patented by Sir John Lawes (1814–1900), fueled the Green Revolution but their environmental dangers were highlighted in the 1960s by American Rachel Carson (1907–1964) in her book *Silent Spring*.

Satellite technology was adopted from the 1960s to monitor the global environment and made a huge impact on meteorology.

Being able to observe the planet from afar also increased worries over its finite nature. In the 1970s Briton James Lovelock argued for the interrelatedness of biological and non-biological factors in a single self-regulating global organism which he called Gaia, amid mounting scientific and popular concern over global pollution and other forms of human impact on the environment. Policy-makers responded with initiatives such as the World Conservation Strategy (1980) and the Earth Summit (1992), which tried to develop new ways of using the planet sustainably.

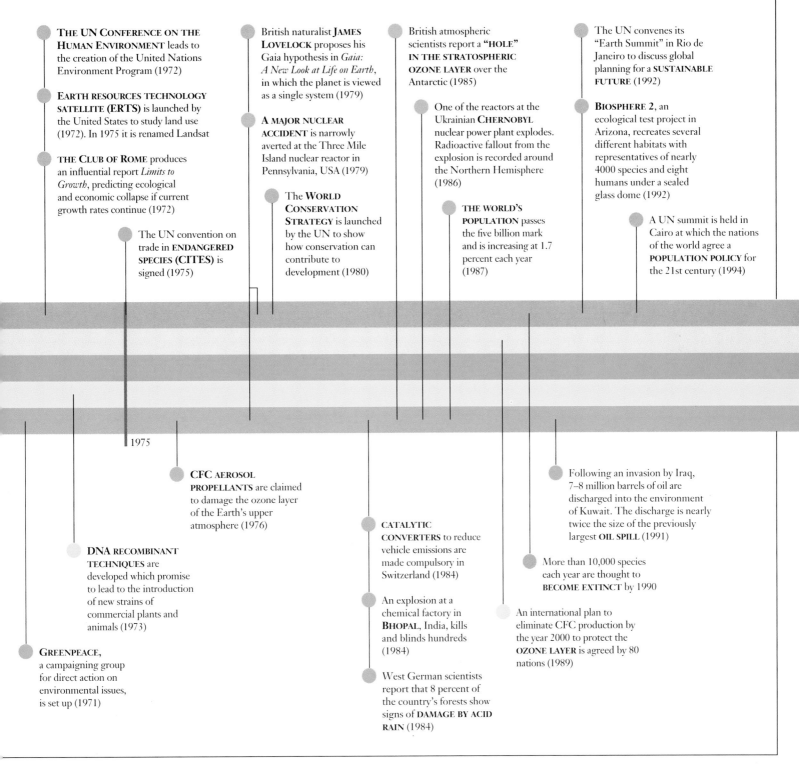

THE UN CONFERENCE ON THE HUMAN ENVIRONMENT leads to the creation of the United Nations Environment Program (1972)

EARTH RESOURCES TECHNOLOGY SATELLITE (ERTS) is launched by the United States to study land use (1972). In 1975 it is renamed Landsat

THE CLUB OF ROME produces an influential report *Limits to Growth*, predicting ecological and economic collapse if current growth rates continue (1972)

The UN convention on trade in **ENDANGERED SPECIES (CITES)** is signed (1975)

British naturalist **JAMES LOVELOCK** proposes his Gaia hypothesis in *Gaia: A New Look at Life on Earth*, in which the planet is viewed as a single system (1979)

A MAJOR NUCLEAR ACCIDENT is narrowly averted at the Three Mile Island nuclear reactor in Pennsylvania, USA (1979)

The **WORLD CONSERVATION STRATEGY** is launched by the UN to show how conservation can contribute to development (1980)

British atmospheric scientists report a **"HOLE" IN THE STRATOSPHERIC OZONE LAYER** over the Antarctic (1985)

One of the reactors at the Ukrainian **CHERNOBYL** nuclear power plant explodes. Radioactive fallout from the explosion is recorded around the Northern Hemisphere (1986)

THE WORLD'S POPULATION passes the five billion mark and is increasing at 1.7 percent each year (1987)

The UN convenes its "Earth Summit" in Rio de Janeiro to discuss global planning for a **SUSTAINABLE FUTURE** (1992)

BIOSPHERE 2, an ecological test project in Arizona, recreates several different habitats with representatives of nearly 4000 species and eight humans under a sealed glass dome (1992)

A UN summit is held in Cairo at which the nations of the world agree a **POPULATION POLICY** for the 21st century (1994)

1975

CFC AEROSOL PROPELLANTS are claimed to damage the ozone layer of the Earth's upper atmosphere (1976)

DNA RECOMBINANT TECHNIQUES are developed which promise to lead to the introduction of new strains of commercial plants and animals (1973)

GREENPEACE, a campaigning group for direct action on environmental issues, is set up (1971)

CATALYTIC CONVERTERS to reduce vehicle emissions are made compulsory in Switzerland (1984)

An explosion at a chemical factory in **BHOPAL**, India, kills and blinds hundreds (1984)

West German scientists report that 8 percent of the country's forests show signs of **DAMAGE BY ACID RAIN** (1984)

Following an invasion by Iraq, 7–8 million barrels of oil are discharged into the environment of Kuwait. The discharge is nearly twice the size of the previously largest **OIL SPILL** (1991)

More than 10,000 species each year are thought to **BECOME EXTINCT** by 1990

An international plan to eliminate CFC production by the year 2000 to protect the **OZONE LAYER** is agreed by 80 nations (1989)

Ecology KEYWORDS

abiotic

Describing a nonliving or physical factor that affects life within an ecosystem. Examples include temperature, light intensity, rainfall and soil structure. Abiotic factors can be harmful to the environment, as when sulfur dioxide emissions from power stations produce acid rain. *See also* **biotic**.

accumulation

The buildup of a persistent **pesticide** within the environment. The accumulation can give rise to genetically resistant pest populations, against which a particular pesticide is no longer effective. *See* **lethal dose** and **persistent**. The term is also used to describe the buildup of increased concentrations of harmful substances (such as pesticides) in the higher levels of the food chain. Producers (plants and photosynthetic microorganisms) and primary consumers (herbivores) may have only low levels of a particular pesticide while carnivores higher up in the food chain may accumulate substantial concentrations.

acid

A substance that, in water, releases hydrogen ions. In modern chemistry, acids are defined as substances that are proton donors and accept electrons to form ionic bonds. Acids react with bases to form salts and also act as solvents. Acids can be detected by using colored indicators such as litmus paper and methyl orange. The strength of an acid is measured by its hydrogen ion concentration, indicated by the pH value.

acid rain

Precipitation (including snow, sleet and fog as well as rain) of an acidity below about pH 4.5–5, principally due to the release of sulfur dioxide and oxides of nitrogen into the atmosphere. Sulfur dioxide is formed by the burning of fossil fuels with high quantities of sulfur. Nitrogen oxides are contributed from various industrial activities and from automobile exhaust fumes. In the atmosphere, sulfur and nitrogenous oxides are converted into sulfuric and nitrogen acids that fall as corrosive rain. Acid rain is linked with damage to and death of forests and lake organisms in northern Europe and eastern North America, as well as industrial areas of Asia. Acid runoff afects the solubility of minerals in soils, lakes and rivers, and can result in the release of toxic substances into solution.

CONNECTIONS

DISRUPTING THE CARBON CYCLE **84**

WATER FACTORS **94**

ATMOSPHERIC POLLUTION **132**

---

adaptation

The slow process of change in the structure or function of an organism that allows it to survive and reproduce more effectively in a particular environment. In evolution, adaptation is thought to occur as a result of random variation in the genetic make-up of organisms coupled with **natural selection**. Species become extinct when they are no longer adapted to their environment; for instance, if the climate suddenly becomes colder. *See* **evolutionary biology**.

Agricultural Revolution

Also known as the Agrarian Revolution, the period in history begining in Britain in the middle of the 18th century, that saw a great change in farming methods and requirements. Smaller and more economic farms were created to meet the increasing demand for food by the growing populations of the cities just before the **Industrial Revolution**. More land was brought under cultivation and enclosed, leading to habitat loss.

agriculture

The practice of farming, including the cultivation of the soil for raising crops and the rearing of domesticated animals. Crops are for human nourishment, animal fodder or commodities such as cotton and coffee. Animals are reared for wool, milk, leather, dung (as fuel and fertilizer), meat or as working animals. *See also* **hunter–gatherer**, **slash-and-burn** and **intensive farming**.

agrochemical

Any synthetic chemical used in modern, intensive agricultural systems. Agrochemicals include nitrate and phosphate fertilizers, pesticides, some animal-feed additives, hormones, growth regulators and various pharmaceuticals. Some agrochemicals are responsible for pollution; for example, the runoff of fertilizers from fields into

,rivers and lakes pollutes the water and destroys marine life (*see* **eutrophication**).

agroforestry

A type of farming in which arable crops are grown on the same land as trees, which are eventually harvested for timber.

air

The mixture of gases that forms the Earth's atmosphere. Dry air has a density of 1.226 kilograms per cubic meter, and consists of about 78 percent nitrogen, 21 percent oxygen, 0.93 percent argon and 0.04 percent carbon dioxide, with small quantities of other inert gases, ozone and water vapor. Air in industrial areas may also be polluted by substantial amounts of sulfur and nitrogen oxides released by factories.

albedo

That fraction of incoming light reflected by an object or material. Its value is always less than or equal to one. Smooth light-colored materials, such as clouds and snow, have a high albedo, close to one, because most of the energy of the incident light is reflected or scattered. Vegetation and large bodies of water have a low albedo, nearer to zero, because they absorb most of the incident energy of the light.

algae

A large group of plantlike organisms that contain the photosynthetic pigment chlorophyll. Ranging from single-celled forms to larhe, complex multicellular seaweeds, they are found mainly in water and damp places. Marine algae help combat **global warming** by removing carbon dioxide from the atmosphere during photosynthesis, and, in some species, fixing it into calcium carbonate structures (limestone) which sink to the ocean floor after the death of the alga. Planktonic algae form the basis of marine and freshwater food chains. Algae used to be included in the division Thallophyta, together with fungi and bacteria, but are now considered to belong to the kingdom Protoctista. *See also* **plankton**.

algal bloom

A sudden increase in the amount of algae in a river or lake, stimulated by the input of nutrients such as phosphates and nitrates. The algae multiply and cover the surface, smothering plants and reducing the light intensity in the upper water layer. The algae eventually die; as their remains are decomposed by bacteria and other microorganisms, oxygen in the water is used up, and fish and other large animals in the water may suffocate. Decaying algae may also give off toxins that

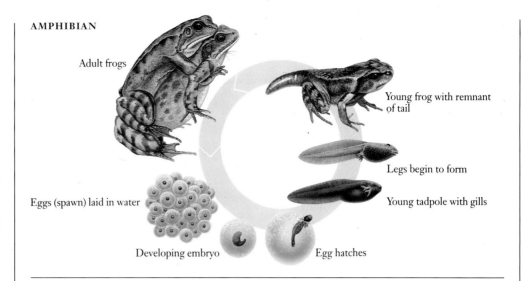

AMPHIBIAN

Adult frogs

Young frog with remnant of tail

Legs begin to form

Eggs (spawn) laid in water

Young tadpole with gills

Developing embryo

Egg hatches

kill fish and other wildlide. The toxins can be harmful to domestic animals and people.

CONNECTIONS

PRIMARY PRODUCERS **72**

THE NITROGEN CYCLE **86**

WATER POLLUTION **134**

alkali

A compound classed as a base that dissolves in water to give hydroxide ions. The four main alkalis are sodium hydroxide, potassium hydroxide, calcium hydroxide and aqueous ammonia. Alkalis react with acids to form a salt and water in a process known as neutralization. The term is also applied to rocks and soils that contain alkali metals.

allelle

One of the alternative forms of a gene. In diploid cells, each gene is present as two alleles, one of which is dominant (expressed), and the other recessive (not expressed).

allopatric

Describing two populations or species that cannot interbreed because the regions they occupy are too far apart.

alpine

Describing the climate typical of those parts of a mountain or mountain range that lie above the tree line but below the permanent snow line. The chief vegetation is grass and other low-growing plants.

alternative energy

Energy produced from renewable and ecologically safe sources, as opposed to nonrenewable sources with toxic byproducts, such as fossil fuels (in conventional power stations) and uranium (in nuclear power sta-

tions). The most important alternative energy source is flowing water, which is harnessed as hydroelectric power. Other sources include waves and tides, wind, the Sun and the heat in the Earth's crust.

CONNECTIONS

CYCLES, CHAINS AND WEBS **68**

ENERGY TRANSFORMATION **70**

NATURAL CYCLES **78**

DISRUPTING THE CARBON CYCLE **84**

ENDLESS ENERGY **88**

amphibian

Any member of the vertebrate class Amphibia, which generally spend their larval (tadpole) stage in fresh water, transferring to land at maturity (after metamorphosis) and usually returning to water to breed. Like fish and reptiles, they continue to grow throughout their lives, and cannot maintain a temperature greatly differing from that of their environment. The class includes the wormlike caecilians, salamanders, frogs and toads.

annual

A plant that completes its life cycle within one year, in which it germinates, matures, flowers, produces seed and dies. Examples include the common poppy and groundsel. Among garden plants, some that are described as annuals are actually perennials (with a life cycle of more than two years), although they are usually cultivated as annuals because they cannot survive winter frosts.

Antarctic

The Earth's southern polar region, extending from the South Pole to the Antarctic Circle, made up of the continent of Antarctica and its surrounding oceans. It has permanent daylight in midsummer and permanent darkness in midwinter. The Antarctic is almost entirely covered by a vast ice sheet and is extremely cold. Most creatures live in the comparatively warmer sea.

Antarctic Circle

The line of latitude located at 66° 33' south of the Equator, representing the northern limit of the Antarctic.

anticyclone

A system of high atmospheric pressure caused by descending air, which becomes warm and dry. Winds radiate from a calm center, taking a clockwise direction in the Northern Hemisphere and an anticlockwise direction in the Southern Hemisphere (*see* **Coriolis effect**). Anticyclones are characterized by calm, fine weather, although in winter, fog and low cloud may also occur.

aquatic

Describing any organism that grows, lives or is found in water. The aquatic environment has a number of advantages. Dehydration is almost impossible, temperatures usually remain stable and the density of water provides physical support.

Arctic

The Earth's northern polar region, extending from the North Pole to the Arctic Circle. It has permanent daylight in midsummer and permanent darkness in midwinter. The area around the Pole consists of a large floating ice cap which in winter extends as far as northern Canada and Russia as the Arctic Ocean freezes over.

Arctic Circle

The line of latitude located 66° 33' north of the Equator, corresponding to the southern limit of the **Arctic.**

arid

Describing a region that is very dry and has little vegetation. Aridity depends on a combination of factors including temperature, rainfall and evaporation. An arid area is usually defined as one that receives less than 250 millimeters of rainfall each year. There are arid regions in northern Africa, Pakistan, Australia, the United States and elsewhere.

arthropod

Any member of the phylum Arthropoda; an invertebrate animal with jointed legs and a segmented body with a horny or chitinous casing (exoskeleton), which is shed periodically and replaced as the animal grows. Included in the phylum are arachnids, such as spiders and mites, as well as crustaceans, millipedes, centipedes and insects.

ANTICYCLONE

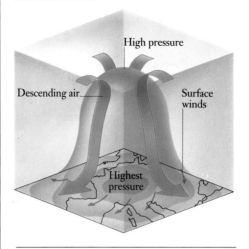

High pressure

Descending air

Surface winds

Highest pressure

artificial classification

A method of classification of organisms that is based on superficial characteristics rather than species interrelationships; for example, the classification of plants into trees, shrubs and herbs. It is considered scientifically inadequate and has been superseded.

asexual reproduction

See **reproduction**.

atmosphere

The envelope of a mixture of gases that surrounds the Earth. It is prevented from escaping back into space by the pull of the Earth's gravity. In its lowest layer (called the troposphere), the atmosphere consists mainly of molecular nitrogen and oxygen. The remainder is largely argon, with very small quantities of other gases, including water vapor and carbon dioxide (*see* **air**). The atmosphere plays a major part in the cycles of nature (*see* **carbon cycle** and **nitrogen cycle**). It is heated by the Earth, which is warmed by infrared and visible radiation from the Sun. Warm air cools as it rises, causing rain and other **precipitation**. Differences in atmospheric pressure between bodies of warm and cold air, together with forces generated by the rotation of the Earth, give rise to the great wind and **weather** systems of the planet. The upper levels of the atmosphere, particularly the ozone layer, absorb almost all of the ultraviolet light radiated by the Sun and prevent lethal amounts of this radiation from reaching the Earth's surface (*see* **ozone**).

CONNECTIONS

THE GLOBAL CLIMATE **50**

DISRUPTING THE CARBON CYCLE **84**

ATMOSPHERIC POLLUTION **132**

---

atmospheric pollution

Contamination of the atmosphere caused by the discharge of a wide range of toxic airborne substances. Often the amount of the released substance is higher in one area, such as surrounding a factory, so the effects are more noticeable. The cost to industry of preventing discharge into the air is very high, so attempts are more usually made to reduce it gradually and disperse it as quickly as possible by using a very tall chimney or by intermittent release. Motor vehicles may be fitted with anti-pollution devices such as catalytic converters, and burn unleaded fuel.

autecology

The study of the relationship between a single species and its environment. *See also* **ecology** and **synecology**.

autotrophic

Describing organisms that produce their own organic material (food) directly from inorganic compounds, using oxidation processes or energy from light. Plants that contain chlorophyll are autotrophs, which produce organic compounds from simple inorganic compounds, such as carbon dioxide and water, during photosynthesis. Some bacteria are also autotrophic. Those that contain bacteriochlorophyll are phototrophic; others, such as those that obtain energy by oxidizing hydrogen sulfide, are said to be chemotrophic. *See also* **heterotrophic**.

bacteria

A group of microscopic unicellular organisms. They usually reproduce by binary fission (dividing into two), and because this may occur approximately every 20 minutes, a single bacterium is potentially capable of producing 16 million copies of itself in a day.

Bacteria can be classified into two broad classes (Gram positive and Gram negative) by their reactions to certain stains, or dyes, used in microscopy. The staining technique, called the Gram test after Danish bacteriologist Hans Gram, allows microbiologists to identify bacteria quickly.

Many bacteria are parasites that cause animal diseases, such as bubonic plague, tuberculosis, cholera and tetanus, and plant diseases, such as soft rot. However, certain types of bacteria are vital in many food and industrial processes, while others play essential roles in the decomposition of organic matter (*see* **carbon cycle** and **nitrogen cycle**). Bacteria cannot normally survive temperatures above 100°C, such as those produced in cooking or pasteurization; but some live around hot vents in the ocean floor of the eastern Pacific and are believed to withstand temperatures of 350°C.

beach

A strip of land bordering the sea, normally consisting of boulders and pebbles on exposed coasts or sand on sheltered coasts. It is usually defined by the high and low tide marks. A berm, a ridge of sand and pebbles, may be found at the farthest point that the seawater reaches.

The material of the beach consists of a rocky debris eroded from exposed rocks and headlands. The material is transported to the beach, and along the beach, by waves that hit the coastline at an angle, resulting in a net movement of the material in one particular direction. This movement is known as longshore drift. Attempts are often made to restrict longshore drift by erecting barriers (breakwaters or groynes) at right angles to the movement. Beaches may also be threatened by the commercial extraction of sand and gravel, by the mineral industry and by pollution (for example, by oil spilled or dumped at sea).

behavior

The pattern of activities by which organisms, particularly animals, react to certain stimuli. Behavioral activities include searching for food, locating a mate (*see* **courtship**), parental care of offspring and avoiding predators. The study of the behavior of animals in their natural environment is known as ethology.

benthic

Describing the region at the bottom of a pond, lake or **ocean**, and the animals that live there. *See also* **pelagic**.

biodegradable

Describing a substance that can be broken down by living organisms (principally bacteria and fungi) into its simpler substances. In organic wastes such as rotting food and sewage, for example, the natural processes of decay lead to compaction and liquefaction, and to the release of nutrients that are then recycled by the ecosystem. Nonbiodegradable substances, such as glass, heavy metals and most types of plastic, present serious problems of disposal.

biodiversity

An overall measure of the variety of the Earth's animal, plant and microbial species, of genetic differences within species, and of the ecosystems that support those species. The maintenance of biodiversity is important for ecological stability and as a resource for research into, for example, new drugs and crops. In the 20th century, the destruction of habitats (for instance, through the spread of agriculture and the cutting down of rainforests) is believed to have resulted in the most severe and rapid loss of diversity in the history of the planet.

CONNECTIONS

ADAPTATION AND EVOLUTION 104

CONSERVING AND RESTORING 140

PRESERVING DIVERSITY 142

---

biological control

The control of pests by living organisms, such as insects and fungi, rather than by the use of chemicals. Methods of biological control include breeding resistant crop strains, inducing sterility in the pest, infecting the pest species with disease organisms or introducing the pest's natural predator. Biological control tends to be naturally self-regulating; however, as ecosystems are so complex, it is difficult to predict all the consequences of introducing a biological controlling agent. *See* **integrated control**.

biological oxygen demand (BOD)

The amount of dissolved oxygen taken up by microorganisms in a sample of water in order to oxidize fully any organic material in the sample. Because these microorganisms live by decomposing organic matter, and the amount of oxygen used is proportional to their number and metabolic rate, the BOD can be used as a measure of the extent to which the water is polluted with organic compounds.

biomass

The total mass of living organisms in a given population or **trophic level**. It may be specified for a particular species or for a general category. Estimates also exist for the entire global plant biomass. Measurements of biomass can be used to study interactions between organisms, the stability of those interactions and variations in population numbers. The burning of biomass (forests, grasslands and fuelwoods) accounts for up to 40 percent of the world's annual carbon dioxide production.

biome

A broad ecological community of plants and animals shaped by common patterns of vegetation and climate. A biome forms the largest land community recognized by ecologists. Examples include tundra, savanna, grassland, desert, and temperate and tropical rainforest. *See also* **habitat**.

biosphere

The part of the Earth that supports life. It is limited to the waters of the Earth, a fraction of its crust and the lower regions of the atmosphere. The biosphere contains a number of different **habitats**.

biotechnology

The use of living organisms to manufacture food, drugs or other products on an industrial scale. The brewing and baking industries have long relied on yeast microorganisms for fermentation, and the dairy industry employs a range of bacteria and fungi to convert milk into cheeses and yogurts. Enzymes, whether extracted from cells or produced artificially, and the biochemical processes they catalyze are central to most biotechnological applications. Recent advances include **genetic engineering**, in which single-celled organisms with modified DNA are used to produce insulin and other drugs.

biotic

Describing influences on the environment that result from the activities of living organisms. Biotic factors include competition and predation. *See also* **abiotic**.

bird

A backboned animal of the class Aves, the biggest group of land vertebrates, characterized by warmbloodedness, feathers, wings, breathing through lungs and egg-laying by the female. There are nearly 8,500 species. Birds are bipedal, with the front limbs modified to form wings and each retaining only three digits. Most birds fly, but some groups (for example, kiwis and ostriches) are flightless, and others include flightless members. Many communicate by sounds, or by visual displays, in connection with which many species are brightly colored (usually only the males). Birds have highly developed patterns of instinctive behavior. Hearing and eyesight are well developed, but the sense of smell is usually poor. Typically the eggs are brooded in a nest and, on hatching, the young receive a period of parental care.

birth rate

The number of live births per thousand of the population over a period of time, usually a year (sometimes it is also expressed as a percentage). For example, a birth rate of 20 per thousand (or 2 percent) means that 20 offspring were born per thousand of the population. Birth rate is a major factor in the study of population change.

CONNECTIONS

POPULATION CURVES 110

POPULATION OUT OF CONTROL 114

---

blue–green algae

The former name for **blue-green bacteria**.

blue-green bacteria

Single-celled primitive organisms, also known as cyanobacteria, that resemble bacteria in their internal cell organization. They are sometimes joined in colonies or filaments. Blue–green bacteria are among the oldest known living things and belong to the kingdom Monera. They are widely distributed in aquatic habitats, on the damp surfaces of rocks and trees, and in the soil. Some can fix nitrogen (see **nitrogen fixation**) and so are necessary to the passage of nitrogen through the environment (see **nitrogen cycle**), while others follow a symbiotic existence; for example, living in association with fungi to form lichens. In some cases fresh water can become polluted by nitrates and phosphates from fertilizers and detergents, and this overenrichment (**eutrophication**) of the water causes the blue–green bacteria to multiply and form **algal blooms**.

boreal forest

Also called taiga, the northern coniferous forests, which form a belt between latitudes 55° and 70° N. Boreal forest may also occur on mountain slopes farther south. The largest area of boreal forest extends from Scandinavia across northern Eurasia. The North American boreal forest is less extensive but has a wider range of species.

bulb

An underground stem with a bud consisting of a compact whorl of fleshy leaves. Some plants overwinter as bulbs and sprout again the following spring.

cacti

Specifically, any member of the plant family Cactaceae, although the term is commonly

BULB

Bud

Fleshy leaves

Stem

Adventitious roots

applied to many different succulent and prickly plants. True cacti have a woody axis (central core) overlaid with an enlarged fleshy stem, which assumes various forms and is usually covered with spines (reduced leaves). They all have adaptations to growing in dry areas, such as reduced transpiration.

calcicole

Any plant that grows best on alkaline soils (soils that are rich in calcium in the form of chalk and limestone). Such plants are also known as calciphiles. See also **calcifuge**.

calcifuge

Any plant that grows best on acid soils. Such plants are also known as acidophiles. See also **calcicole**.

canopy

The upper level of mature woodland formed by the highest branches and foliage of trees. The level below is the shrub level.

captive

Describing any organism that is confined or restrained. Captive breeding programs are often the last resort of conservationists trying to save a single species from extinction. See also **zoo**.

carbohydrate

A member of a group of chemical compounds composed of carbon, hydrogen and oxygen with the basic formula $C_m(H_2O)_n$ and related compounds with the same basic structure but modified functional groups. The simplest carbohydrates are sugars (monosaccharides, such as glucose and fructose, and disaccharides, such as sucrose). When these basic sugar units are joined together in long chains or branching structures they form polysaccharides, such as starch and glycogen, which often serve as food stores. Cellulose is the chief structural carbohydrate in plants. Carbohydrates are produced in autotrophs (plants and some microorganisms) by photosynthesis and, as sugar and starch, they are major energy providers in animal nutrition (see **producer**). Animals geneate their own carbohydrate stores by converting sugars to the insoluble polysaccharide glycogen.

carbon cycle

The sequence of chemical reactions by which carbon circulates and is recycled through the ecosystem. Carbon from carbon dioxide is taken up by plants during photosynthesis and converted into carbohydrates, releasing oxygen into the atmosphere. The carbohydrates are then used directly by the plant – or by animals that eat plants – in

respiration, in which they are oxidized to release carbon dioxide back into the atmosphere. Carbon dioxide is also released into the atmosphere by the burning of fossil fuels. Today, the carbon cycle is in danger of being disrupted by the increased consumption and burning of fossil fuels and the burning of large tracts of tropical forest, as a result of which levels of carbon dioxide are building up in the atmosphere and probably contributing to the **greenhouse effect**.

CONNECTIONS

ENERGY TRANSFORMATION **70**

NATURAL CYCLES **78**

THE CARBON CYCLE **82**

DISRUPTING THE CARBON CYCLE **84**

---

carbon dioxide

A colorless gas, slightly soluble in water and denser than air. It is a waste product produced during respiration in plants, animals and microorganisms, and through the decay of organic matter. However, it is also used in green plants in photosynthesis and is thus a major contribution to the food of virtually all life forms.

carnivore

A flesh-eating mammal of the order Carnivora. There are some 230 species of carnivores, divided into seven families: cats, viverrids (civets and relatives), hyenas, canids (dogs and relatives), bears, procyonids

CARBOHYDRATE

Cellulose

Starch

Hydrogen

Oxygen

Carbon

(raccoons and relatives) and mustelids (weasels and relatives). Carnivores occupy a wide range of habitats and are distributed worldwide. The term is often used more broadly to include any animal that eats other animals, even microscopic ones. *See also* **herbivore** and **omnivore**.

carrion feeder
Any organism that feeds on dead or rotting flesh (carrion). Vultures are carrion feeders.

carrying capacity
The maximum number of animals of a given species that a particular area can support during the harshest part of the year, or the maximum biomass it can sustain indefinitely. When the carrying capacity is exceeded, there is insufficient food (or other resources) for the members of the population. Their numbers may then be reduced by emigration, reproductive failure or death through starvation. *See* **population dynamics**.

catalytic converter
A device fitted to the exhaust systems of an automobile to reduce the amount of toxic emissions. A mixture of catalysts, usually based on platinum, thinly coated on a metal or ceramic honeycomb (to maximize the available surface area), is used to convert harmful substances in the passing exhaust gas into less harmful ones: for example, oxidation catalysts assist the conversion of hydrocarbons in unburned fuel and carbon dioxide into carbon dioxide and water. Three-way catalysts convert oxides of nitrogen back into nitrogen. Catalysts are "poisoned" by lead and sulfur compounds, and these must not be present in the fuel used by automobiles fitted with catalytic converters.

catchment area
A usually hilly or mountainous area which collects rainfall and channels it into a specific system of streams and rivers.

cell
The basic structural and functional unit of all living things. All organisms, with the exception of viruses, consist of one or more cells. Bacteria, protozoa and many other microorganisms consist of single cells, whereas a human is made up of billions of cells. Essential features of a cell are the membrane, which encloses it and restricts the flow of substances in and out; the jellylike material within (cytoplasm); the ribosomes, which carry out protein synthesis; and the **DNA**, which forms the hereditary material. The composition of the protoplasm (cytoplasm and organelles) varies, but the products of its breakdown when the cell dies are

mostly proteins. All cells require nutrients to grow, and they may be damaged by external factors such as pollution.

cellulose
A complex **carbohydrate** composed of long chains of glucose units. It is the principal constituent of the cell walls of higher plants and a vital ingredient in the diet of many herbivores. Molecules of cellulose are organized into long, unbranched microfibrils that give support to the cell wall.

cereal
Any grass grown for its edible, nutrient-rich, starchy seeds. The term refers primarily to wheat, oats, rye and barley, but may include maize (corn), millet and rice. Cereals contain about 75 percent complex carbohydrates and 10 percent protein, together with fats and fiber (roughage). They are easily stored. If all the world's cereal crop were consumed as wholegrain products directly by humans, everyone could obtain adequate protein and carbohydrates; however, a large proportion of cereal production in affluent nations is used as animal feed.

chlorofluorocarbons (CFCs)
A group of synthetic odorless, nontoxic, nonflammable and chemically inert gases containing chlorine, fluorine, carbon and sometimes hydrogen. CFCs have been used as propellants in aerosol cans, in refrigerators and air conditioners, and in the manufacture of foam packaging. When CFCs are released into the atmosphere, they drift up slowly into the stratosphere where, under the influence of ultraviolet radiation from the Sun, they break down into chlorine atoms which break down ozone molecules. This reduction in the ozone layer allows more of the harmful radiation from the Sun to reach the Earth's surface.

CONNECTIONS

THE CARBON CYCLE **82**

DISRUPTING THE CARBON CYCLE **84**

ATMOSPHERIC POLLUTION **132**

---

chlorophyll
The green pigment present in most plants and responsible for the absorption of light energy during **photosynthesis**. The pigment absorbs the red and blue–violet parts of sunlight but reflects the green. Chlorophyll is found within chloroplasts, present in large numbers in leaves. **Cyanobacteria** and other photosynthetic bacteria also have chlorophyll, though of a different type (bacteriochlorophyll).

chromosome
A threadlike structure in the cell nucleus that carries the genetic information (*see* **gene**). Each chromosome consists of a strand of highly folded, coiled DNA and associated proteins. The point on a chromosome where a particular gene occurs is known as its locus. The cells of most higher organisms (excluding the gametes) contain two copies of each chromosome (they are diploid), although some cells contain only one copy (haploid).

clay
A very fine-grained sedimentary deposit with a particle size less than 0.002 millimeters across that has undergone consolidation. When moistened it becomes unstable, making it prone to mass movement. When dry, it hardens to form an impermeable layer. Clays may be white, gray, red, yellow, blue or black in color, depending on the actual composition. Clay minerals consist largely of hydrous silicates of aluminum and magnesium with iron, potassium, sodium and organic substances.

climate
The characteristic **weather** conditions of a particular place over a period of time. It encompasses meteorological elements and the factors that influence them. The primary factors that determine variations of climate over the surface of the Earth are latitude and the tilt of the Earth's axis to the plane of the orbit about the Sun (66.5°); the largescale movements of different wind belts over the

---

CHROMOSOME

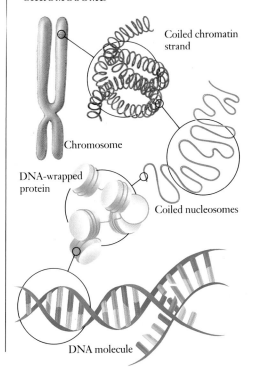

Coiled chromatin strand

Chromosome

DNA-wrapped protein

Coiled nucleosomes

DNA molecule

Earth's surface; the temperature difference between land and sea; contours of the land, and location of the area in relation to ocean currents. Catastrophic variations to climate may be caused by the impact of another planetary body or by clouds resulting from volcanic activity. The most important meteorological changes brought about by human activity are those linked with **ozone** depleters and the **greenhouse effect**. How much heat the Earth receives from the Sun varies in different latitudes and at different times of the year. In the equatorial region the mean daily temperature of the air near the ground has no large seasonal variation. In the polar regions the temperature in the long winter, when there is no incoming solar radiation, falls far below the summer value.

The temperature of the sea, and of the air above it, varies little in the course of day or night, whereas the surface of the land is rapidly cooled by lack of solar radiation. In the same way, the annual change of temperature is relatively small over the sea and great over the land. Continental areas are thus colder than the sea in winter and warmer in summer. Winds that blow from the sea are warm in winter and cool in summer, whereas winds from the central parts of continents are hot in summer and cold in winter. Air temperature drops with increasing land height, thus places situated above mean sea level usually have lower temperatures than places at or near sea level. Even in equatorial regions, high mountains are snow-covered during the whole year. The complexity of the distribution of land and sea, and the consequent complexity of the general circulation of the atmosphere, have a direct effect on the distribution of the climate.

climax community

A group of organisms that is in equilibrium with existing environmental conditions and forms the last stage in the natural **succession**. A mature woodland is an example.

cloud forest

A type of forest that grows at high altitudes on tropical mountains, which receives constant moisture from water condensing out of the air as clouds.

codominant

In genetics, the failure of a pair of alleles (*see* **gene**), controlling a particular characteristic, to show the classic recessive–dominant relationship. Instead, aspects of both alleles may show in the phenotype (visible traits). The snapdragon shows codominance in respect to color. Two alleles, one for red petals and the other for white, produce a pink color if the alleles occur as a **heterozygous** form.

colonization

The spread of species into a new habitat, such as a freshly cleared field or a recently flooded valley. The first species to move in are called pioneers, and may establish conditions that allow other animals and plants to arrive (for example, by improving the condition of the soil or by providing shade). Over time a range of species arrives and the habitat matures; early colonizers are probably replaced, so that the variety of animal and plant life changes. *See also* **succession**.

CONNECTIONS

HABITATS AND NICHES **64**

SUCCESSION **100**

ADAPTATION AND EVOLUTION **104**

COLONIZATION STRATEGIES **114**

---

combustion

Burning, defined in chemical terms as the rapid combination of a substance with oxygen, accompanied by the evolution of heat (exothermic) and usually light. A slow-burning candle flame and the explosion of a mixture of petrol vapor and air are extreme examples of combustion. The complete combustion of carbon-containing fuels results in the production of carbon dioxide and water vapor. Incomplete combustion occurs when a substance has too little oxygen to burn completely. Most combustion processes are carried out to obtain the heat from combustion. However, some combustion processes are used to produce specific products (for example, burning sulfur in air gives sulfur dioxide for sulfuric acid production).

community

A naturally occurring assemblage of plants and animals that occupy a common environment. Communities are usually named after a dominant feature such as a characteristic plant species (for example, beechwood).

competitive exclusion principle

Also known as Gause's principle, a law that states that two species having identical ecological requirements cannot coexist in the same habitat unless there is a superabundance of environmental resources, such as food, in that particular habitat.

compost

Organic material, such as vegetable or municipal wastes, decomposed by bacteria under controlled conditions to make a nutrient-rich natural fertilizer. A well-made compost heap reaches a high temperature during the composting process, killing most weed seeds that might be present. *See also* **humus**.

compound

In chemistry, any substance made up of two or more elements bonded together, so that they cannot be separated by physical means. The formation of a compound requires a chemical reaction. Compounds are held together by ionic or covalent bonds. In botany, the term compound describes a leaf made up of a number of smaller leaflets.

condensation

The change of phase of a vapor to a liquid. Condensation usually occurs when the vapor comes into contact with a cold surface. It is an essential step in distillation processes and is also the process by which water vapor high in the air turns into fine water droplets to form clouds. Condensation in the atmosphere occurs when the air becomes completely saturated and is unable to hold any more water vapor. As air rises it cools and contracts – the cooler it becomes the less water it can hold. The temperature at which the air becomes saturated is known as the dew point. Condensation is an important element of the recycling of water in the environment.

cone

A compact group of reproductive organs found in conifers and cycads, also known as a strobilus. A cone consists of a central axis surrounded by numerous, overlapping, scalelike, modified leaves (sporophylls) that bear the reproductive organs. Usually there are separate male and female cones, the former bearing pollen sacs containing pollen grains, and the larger female cones bearing the ovules that contain the ova or egg cells. The pollen is usually carried from male to female cones by the wind (anemophily). The seeds develop within the female cone and are released as the scales open in dry atmospheric conditions, which favor seed dispersal.

conifer

Any tree or shrub of the class Coniferophyta, in the gymnosperm plant group. They are often pyramidal in form, with leaves that are either scales or made up of needles; most are evergreen. Conifers include pines, spruces, firs, yews, junipers, monkeypuzzles and larches (which are not evergreens). The reproductive organs are the male and female cones, and pollen is generally distributed by the wind. The seeds develop in the female cones. The processes of maturation, fertilization and seed ripening may extend over several years. Most conifers grow quickly and can tolerate poor soil, steep slopes and short growing seasons. They are the main forest trees of colder regions.

coniferous

A cone-bearing plant. Coniferous forests are habitats dominated by cone-bearing plants. *See* **conifer**.

conservation

The planning and management actions taken to protect and preserve the natural world, usually from pollution, overexploitation and other harmful features of human activity. The late 1980s saw a great increase in public concern for the environment, with membership of various conservation groups rising sharply. Globally the most important issues include the depletion of atmospheric ozone by the action of **chlorofluorocarbons** (CFCs), the buildup of carbon dioxide in the atmosphere (thought to contribute to an intensification of the **greenhouse effect**) and **deforestation**.

CONNECTIONS

HABITATS AND NICHES **64**

DISRUPTING THE CARBON CYCLE **84**

OVEREXPLOITATION **128**

CONSERVING AND RESTORING **140**

---

consumer

Any organism whose feeding habits cause it to be classified above the first (**producer**) level in a **food chain**. Primary consumers are usually **herbivores**. Secondary consumers are **carnivores**.

continental shelf

The gently sloping sea floor off the coast of a continent, which may extend as far out as 1200 kilometers from the shore, at a depth up to 370 meters. Beyond this, the shelf gives way to the continental slope and then to deep water. Such waters are typically rich in nutrients and in marine life.

CORIOLIS EFFECT

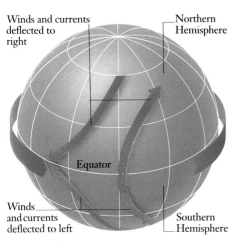

Winds and currents deflected to right

Northern Hemisphere

Equator

Winds and currents deflected to left

Southern Hemisphere

contraception

The use of any drug, device or surgical technique that prevents pregnancy. Human contraceptives in current use include the contraceptive pill, which contains female hormones that interfere with ovulation or the first stage of pregnancy. Barrier contraceptives include condoms and diaphragms that prevent sperm entering the cervix. Intrauterine devices (IUDs) cause a slight inflammation of the uterus lining which prevents the fertilized egg from becoming implanted. Other contraceptive methods in use include sterilization (women) and vasectomy (men); these are usually nonreversible. Contraception is encouraged in some countries as a means of reducing the birth rate and controlling population growth. In other countries it is banned on religious grounds.

convection

The transfer of heat as the result of the movement of a fluid (such as air or water). This may be the result of temperature difference; the warmer parts of the fluid are less dense and therefore tend to rise. The vertical convection of air leads to cooling. Free convection occurs when air, heated by conduction from an underlying landmass, rises. Forced convection occurs when a body of air encounters higher ground (for example, mountains) that lifts the air regardless of the initial temperature. Forced convection can lead to very high rainfall in mountainous regions. *See* **climate**.

coral

Any of a group of marine invertebrates of the class Anthozoa in the phylum Cnidaria, and a few species of the class Hydrozoa (hydroids). Corals secrete a skeleton of calcium carbonate extracted from the surrounding water. They occur in warm seas, at moderate depths with sufficient light. Corals live in a symbiotic (mutually beneficial) relationship with microscopic algae (zooxanthellae), in which the algae obtain carbon dioxide from the coral polyps and the polyps receive nutrients from the algae. Most corals form large colonies, although there are species that live singly. Their accumulated skeletons make up coral reefs and atolls. *See also* **reef**.

Coriolis effect

The effect of the Earth's rotation on the atmosphere and on all objects on the Earth's surface. In the Northern Hemisphere the Coriolis effect causes objects and air currents moving in a north-south direction to be deflected to the right; in the Southern Hemisphere it causes deflection to the left. The effect is named after its discoverer, French mathematician Gaspard Coriolis.

Potato

Barley

Peas and beans

Wheat

1

2

3

Fixed nitrogen

4

The Coriolis effect can be easily observed by watching previously stationary water go down a plughole – it does not flow directly downward but spins to the right (clockwise) or to the left (anticlockwise), depending on whether the observer is in the Northern or Southern Hemisphere.

courtship

Behavior patterns exhibited by animals as a prelude to mating. Such patterns vary considerably from one species to another, but are often ritualized forms of behavior not obviously related to courtship or mating (for example, courtship feeding in birds). Courtship ensures that copulation occurs with a member of the opposite sex of the right species. It also synchronizes the partners' readiness to mate and allows each partner to assess the suitability of the other.

crop rotation

An important method for ensuring continuing soil fertility in cultivated areas in which the crops grown on a particular piece of land are changed regularly on a yearly basis. Crop rotation has been used in Europe and Asia for many centuries. The crops are grown in a particular order to utilize and add to the nutrients in the soil and to prevent the buildup of insect and fungal pests. Including a legume crop (such as peas or beans), in the rotation builds up nitrates in the soil because the roots of legumes contain bacteria which fix nitrogen from the air (see **nitrogen fixation**).

current

The flow of a body of water or air moving in a definite direction. Ocean currents are fast flowing currents of seawater generated by the wind or by variations in water density between two areas. They are partly responsible for transferring heat from the Equator to the Poles and thereby evening out the global heat imbalance. There are three basic types of ocean current: drift currents are broad and slow moving, stream currents are narrow and swift moving, and upwelling currents bring cold, nutrient-rich water from the ocean depths. Stream currents include the Gulf Stream and the Japan (or Kuroshio) current. Upwelling currents, such as the Gulf of Guinea current and the Peru (Humboldt) current, provide food for plankton, which in turn supports fish and sea birds. *See also* **convection**.

cuticle

The hard noncellular protective surface layer of many invertebrates such as insects. In plants, the cuticle is the waxy surface layer on those parts of plants that are exposed to the air. All types of cuticle are secreted by the cells of the epidermis. A cuticle reduces water loss and, in arthropods, forms an exoskeleton.

cyanobacteria

See **blue-green bacteria**.

cycle

A recurrent series of events in which the conditions at the end of the sequence are the same as at the beginning. Important examples of cycles in nature are the **carbon cycle**, **nitrogen cycle**, phosphorus cycle, sulfur cycle and **water cycle**.

> ### CONNECTIONS
> ENERGY TRANSFORMATION **70**
> PYRAMIDS AND WEBS **76**
> NATURAL CYCLES **78**
> THE CARBON CYCLE **82**
> THE NITROGEN CYCLE **86**
> THE URANIUM CYCLE **90**

cyclone

An area of low atmospheric pressure, around which the air circulates. In the Northern Hemisphere, winds blow in a counterclockwise direction; they blow clockwise around cyclones in the Southern Hemisphere. Cyclones are also known as depressions and are associated with storms. Severe cyclones that form in the Tropics are known under various names, such as tropical cyclones, hurricanes or typhoons. *See also* **anticyclone**.

DDT

Dichlorodiphenyltrichloroethane (chemical formula $(ClC_6H_5)_2CHCHCl_2$), a persistent organic insecticide. It was originally used widely in the control of mosquitoes that spread malaria, although resistant strains developed after prolonged use. DDT has had a disruptive effect on species high in the food chain, in particular on the breeding success of certain predatory birds, due to its accumulation in animal tissues. Its use is now banned in most countries.

> ### CONNECTIONS
> PYRAMIDS AND WEBS **76**
> AGROCHEMICALS **122**
> AGROCHEMICALS AND THE BIOSPHERE **124**
> BIOLOGICAL ALTERNATIVES **126**

death rate

The number of deaths per thousand of the population of an area over one year. In humans it is linked to social and economic factors such as standard of living, diet and access to clean water; in animal populations it is most affected by predation, disease and food supply. *See* **population dynamics**.

deciduous

Describes trees and shrubs that shed their leaves at the end of the growing season or during a dry season to reduce the loss of water from the leaves by transpiration (evaporation). Most deciduous trees are

DDT

Fruit tree

Insect pest

Contaminated fruit

Bird of prey

Fruit-eating bird

Some birds die

Bird of prey dies

angiosperms – plants in which the seeds are enclosed within an ovary. For this reason, the term deciduous tree is sometimes used to mean angiosperm tree, although many angiosperms are evergreen (especially in the Tropics), and a few gymnosperms – plants in which the seeds are exposed – are deciduous (for example, larches). The term broadleaved is now preferred.

decomposer

Any organism that breaks down dead organic matter. Decomposers play a vital role in the ecosystem and the food chain by freeing important chemical substances, such as nitrogen compounds, locked up in dead organisms or excreta. They feed on some of the released organic matter, but leave the rest to filter back into the soil or pass in gaseous form into the atmosphere. The principal decomposers are bacteria and fungi, but earthworms and many other invertebrates also perform this function.

decomposition

The chemical processes by which a compound is reduced to its component substances. In biology, it is the destruction of dead organisms either by chemical reduction or by the action of **decomposers**.

defoliation

The removal of leaves, particularly from trees, through the action of drought or pests, or as a deliberate act of war (to rob an enemy of cover) by spraying forests with defoliants. The normal seasonal loss of leaves by **deciduous plants** is also called defoliation.

deforestation

The permanent removal of forest and undergrowth; destruction of forest for timber, fuel, and clearing for agriculture and extraction industries, such as mining, without planting new trees or working on a cycle that allows the natural forest to regenerate. Largescale deforestation causes fertile soil to be blown away or washed into rivers, leading to soil erosion, drought, flooding and loss of wildlife. It may also lead to locally decreasecd rainfall. With fewer trees absorbing carbon dioxide from the air for photosynthesis, deforestation may also increase the carbon dioxide content of the atmosphere and intensify the **greenhouse effect**. *See also* **global warming**.

> ### CONNECTIONS
> DISRUPTING THE CARBON CYCLE **84**
> OVEREXPLOITATION **128**
> DISAPPEARING FORESTS **136**

demography

The study of the dynamics (for example, size, and age and sex structure) of human populations to establish reliable statistics on such factors as birth and death rates, marriage and divorce rates, life expectancy and migration. Demography is used to calculate life tables, which give the life expectancy of members of the population by sex and age. Demography is significant in the social sciences as the basis for industry and for government planning in such areas as education, housing, welfare, transport and taxation. *See* **survivorship curve, birth rate, death rate** and **population dynamics** .

denitrification

The process, occurring naturally in soil, by which bacteria break down nitrates to produce nitrogen gas, which returns to the atmosphere. *See* **nitrogen cycle**.

density-dependent

A situation in which population growth depends on the existing population density. *See* **population dynamics**.

density-independent

A situation in which the percentage mortality or survival of a species varies independently of the actual population density. *See* **population dynamics**.

derelict

Land that has been damaged by industrial processes or neglect and is unfit for any use without rehabilitating treatment.

desert

A very arid area without sufficient rainfall and vegetation to support human life. Almost one-third of the Earth's land surface is desert. The tropical desert belts between latitudes from 5° and 30° are caused by the descent of air that is heated over the warm land and therefore retains its moisture. Other natural desert types are the continental deserts, which are too far from the sea to receive any moisture; rain-shadow deserts, which lie in the lee of mountain ranges, where the ascending air drops its rain only on the windward slopes; and coastal deserts, in which cold ocean currents cause local dry air masses to descend. Deserts can be created by changes in climate or by the human-aided process of **desertification**.

desertification

The creation of deserts caused by climatic changes or by artificial processes. It can sometimes be reversed by planting special vegetation (grasses or trees) and by the use of water-absorbent plastic grains which, added to the soil, enable crops to be grown. The processes leading to desertification include **overgrazing**; the destruction of forest belts; exhaustion of the soil by intensive cultivation without the use of fertilizers; and salinization of soils due to badly managed irrigation. *See also* **deforestation**.

desiccation

Drying out through the loss or unavailability of water, as in a prolonged drought. The deliberate drying out of newly felled timber – seasoning – is also known as desiccation.

detergent

A surface-active agent used to remove dirt and grease. Common detergents are made from fats and sulfuric acid. Phosphates in some detergents can cause the excessive enrichment (**eutrophication**) of rivers and lakes. "Environmentally friendly" detergents contain no phosphates or bleaches.

detrivore

An animal that feeds on detritus, small particles of organic matter produced during the decomposition of animals and plants. Many detrivores live at the bottom of the sea and feed on particles that drift down from above.

dew

Moisture that condenses on solid objects after their temperature has fallen below the dew point of the surrounding air (*see* **condensation**). As the temperature falls during the night, the air and its water vapor become chilled, and condensation takes place on, for example, the ground, which cools more rapidly than the air. If the temperature falls below freezing point during the night, the dew freezes to form hoar **frost**.

diet

The range of foods consumed by an animal. The basic components of a diet are proteins, carbohydrates, fats, vitamins, minerals and water. Different animals require these substances in different proportions, and have digestive systems adapted to cope with their particular diets. For humans, an adequate diet fulfills the body's nutritional requirements and gives an energy intake proportional to activity level. Dietary requirements may vary over an organism's lifespan with growth, reproduction or old age.

disease

Any condition that impairs the normal state of an organism and usually alters the functioning of one or more of its organs or systems. A disease is usually characterized by a set of specific symptoms and signs, although these may not always be apparent to the sufferer. Diseases may be inborn (congenital), or acquired through viral, bacterial or fungal infection, injury or other causes.

distribution of species

The occurrence of a species in geographical location, altitude or some other factor.

diurnal

Occurring on a daily basis or during the daytime. For example, the diurnal cycle of **pollution** is of great interest to those concerned with pollution control. *See also* **nocturnal**.

DNA (deoxyribonucleic acid)

A large complex molecule that contains the chemically coded information to build, control and maintain a living organism. DNA, a ladderlike double-stranded nucleic acid, is the basis of genetic inheritance in virtually all living organisms. In organisms other than bacteria it is organized into chromosomes and contained in the cell nucleus. DNA is made up of two chains of nucleotide subunits, with each nucleotide containing sugar, phosphate and either a purine (adenine or guanine) or pyrimidine (cytosine or thymine) base. The bases link with each other (adenine linking with thymine, and cytosine with guanine) in base pairs to connect the two strands of the DNA molecule's ladderlike "rungs". The sequence of bases forms a code for the information stored in the DNA. *See* **gene**.

CONNECTIONS

ADAPTING TO THE ENVIRONMENT **102**
ADAPTATION AND EVOLUTION **104**
BIOLOGICAL ALTERNATIVES **126**
PRESERVING DIVERSITY **142**

---

doldrums

A calm, windless region occurring in oceans that lie astride the Equator.

domestication

The process by which wild animals or plants are brought under control or cultivation. Domesticated farm animals include cattle, pigs and sheep. All basic crops are descended from originally wild plants that have been domesticated.

dominance

In ecology, the characteristic and often tallest species in a particular plant community. The dominant species exerts the greatest overall influence on the community (for example, oak in oakwood). Dominance also applies to a type of gene in genetics.

ECOSYSTEM

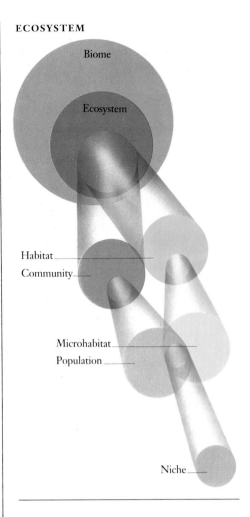

dormancy
A state of lowered metabolism and lack of activity in a plant or animal. The seeds of many plants remain dormant in winter, as do the spores of other organisms. In animals, extended winter dormancy is referred to as **hibernation**. *See also* **estivation**.

drought
A period of unusually low rainfall. In North America and Europe, it may last only a few weeks, whereas in **desert** areas a drought may be a succession of unusually dry years. The areas of the world subject to serious droughts, such as the Sahara, are increasing because of the destruction of forests, overgrazing and poor agricultural practices.

dust storm
A storm of dust blown up when winds are strong enough to lift it from the ground. The critical wind speed depends on the specific gravity, shape, size and dampness of the dust particles. Dust storms may be local – associated with thunderstorms and rain clouds – or extensive, associated with areas of low atmospheric pressure. Dust devils are small storms caused by intense local heating of the ground in desert areas.

ecology
The study of the relationships between living organisms and their environments, including all living and nonliving components. The term was coined by the biologist Ernst Haeckel in 1866. Ecology may be concerned with individual organisms (for example, behavioral ecology and feeding strategies); with populations (population dynamics); or with entire communities (competition between species for access to resources in an **ecosystem** or predator–prey relationships). Applied ecology is concerned with the management and conservation of habitats, and with the consequences and control of pollution. *See* **synecology** and **autecology**.

ecosystem
An integrated ecological unit consisting of the living organisms and the physical environment (**biotic** and **abiotic** factors) in a particular area. Ecosystems can be defined on different scales. The global ecosystem consists of all the organisms living on Earth, the Earth itself (both land and sea) and the atmosphere above. A freshwater pond ecosystem consists of the plants and animals in the pond, the pond water and all the substances dissolved or suspended in that water, and the rocks, mud and decaying matter that make up the pond bottom. Energy and nutrients pass through the ecosystem in a particular sequence (*see* **food chain**).

Ecosystems are fragile because the relationships between their components are so finely balanced: the alteration or removal of any one component can sometimes cause the whole ecosystem to collapse. The removal of a major predator, for example, can result in the destruction of the ecosystem through overgrazing by increased numbers of herbivores. The tropical rainforests of South America, central Africa and Southeast Asia are examples of ecosystems under threat.

ectothermic
Describing an animal whose body temperature varies with the external temperature. Such animals are commonly (but inaccurately) called cold-blooded. Invertebrates, fish, amphibians and reptiles are ectothermic. Birds and mammals are **endothermic** (warm-blooded**)**.

edaphic
Describing the biological, chemical and physical characteristics of the soil that affect a given ecosystem.

effluent
Liquid sewage or industrial waste, partially or completely treated, that flows out of a treatment plant.

El Niño
Originally, a warm current that periodically (about once every 10 years) flows south along the northwestern coast of South America, where the current is normally cold. The warm water kills fish, forcing seabirds to migrate to find food, and affects continental weather patterns. The term is now applied to the more frequent intense and extensive warming of the eastern Pacific, and associated weather changes in the tropics.

emigration
The movement of an individual from one environment to settle permanently in another. *See* **migration** and **population dynamics**.

endangered species
Any plant or animal whose numbers are so few that it is at risk of becoming extinct. Officially designated endangered species are listed by the International Union for the Conservation of Nature (IUCN) in a series of "red data books". An example is the Javan rhinoceros. There are only about 50 alive today and, unless active steps are taken to promote this species' survival, it will probably be extinct within a few decades.

endothermic
Describing an animal whose metabolism generates enough energy to maintain its body temperature within a narrow range. Such animals are commonly (but not correctly) called "warm-blooded." Only birds and mammals are endothermic. Other animals are **ectothermic.**

energy
The capacity for doing work. It has many forms including mechanical, chemical, thermal, nuclear and electrical. Energy resources are stores of convertible energy. Nonrenewable resources include the fossil fuels (coal, oil and gas) and nuclear-fission fuels. Renewable resources, such as wind, tidal and geothermal power, have so far been less exploited than nonrenewable sources. *See* alternative energy.

energy balance
The overall result of energy exchanges between organisms in an ecosystem and their surroundings.

energy efficiency
Preserving energy resources by reducing the use of energy. Profligate and inefficient energy use by industrialized countries contributes greatly to **atmospheric pollution** and the **greenhouse effect**, when it draws on nonrenewable energy sources.

environment

The sum of the **abiotic** and **biotic** conditions that affect a particular organism (*see* **biosphere** and **habitat**). The term is often used to refer also to the total global environment, without reference to any particular organism.

environmental resistance

The restriction of population growth due to environmental factors. For example, increased numbers of grazing animals leads to increased competition for food between them, reducing the food supply, which prevents population growth.

Equator

An imaginary line, 40,092 kilometers long, encircling the broadest part of the Earth, and representing 0° **latitude**. It divides the Earth into two halves, the Northern and Southern Hemispheres.

equilibrium

A stable state; an equilibrium population is neither increasing or decreasing, but fluctuates around the **carrying capacity**, which depends on the viability of the **habitat**. *See also* **population dynamics**.

EROSION

Cave
Headland
Arch
Stack

erosion

The breaking down of solid rock into smaller particles and their transportation. Agents of erosion include water (the sea, rivers), glaciers and wind. The human population also contributes directly to erosion by bad farming practices and the cutting down of forests. *See also* **weathering**.

estivation

A state of **dormancy**, similar to **hibernation**, that occurs during the dry season in animals such as lungfish and snails.

estuarine

Describing an environment formed at a river estuary (that is, where a river mouth widens into the sea), where fresh water mixes with salt water and tidal effects are felt.

eutrophic

Describing freshwater habitats that are rich in plant nutrients. Such areas of water can give rise to shortlived **algal blooms** that may kill aquatic animals and higher plants. *See also* **eutrophication**.

eutrophication

The excessive enrichment of a body of water, primarily by nitrate fertilizers washed from the soil by rain, phosphates from fertilizers and detergents, nutrients in municipal sewage, and sewage itself. Eutrophication leads to an increase in the growth of aquatic plants and algae (*see* **algal blooms**), which compete for light. Some cannot get enough light, and they die. As the dead organisms decompose, oxygen is used up by the decomposers; this leads to suffocation of aquatic animals, because the bloom prevents oxygen replenishment from the atmosphere.

CONNECTIONS

PRIMARY PRODUCERS 72

THE NITROGEN CYCLE 86

THE NITROGEN CYCLE 86

WATER POLLUTION 134

---

evergreen

A plant – especially a tree – that does not shed its leaves in winter (unlike a **deciduous** plant, which does). Conifers and tropical rainforest trees are typical evergreens.

evolution

The slow process of cumulative change from one form to another in successive generations of living organisms, which leads to the development of different species and subspecies from a common ancestor.

evolutionary biology

The study of the theories that try to explain the origins of different species through gradual changes in ancestral groups. The idea gained wide acceptance in the 19th century, following the work of the Scottish geologist Charles Lyell, the French naturalist Jean Baptiste Lamarck, the English naturalists Charles Darwin and Alfred Wallace, and the English biologist Thomas Henry Huxley. Darwin assigned the major role in evolutionary change to **natural selection** acting on randomly occurring variations. The current theory of evolution (neo-Darwinism) combines Darwin's theory with the Austrian biologist Gregor Mendel's theories on genetics and Hugo de Vries's discovery of genetic mutation. In addition to natural and sexual selection, chance may influence which genes become characteristic of a population, a phenomenon called genetic drift.

exotic species

Any non-native species of animal or plant that requires human intervention to survive.

exponential

A mathematical function in which the variable quantity is an exponent (a number indi-

---

EVOLUTION

1 Modern horse (Equus)
2 Equus
3 Merychippus
4 Mesohippus
5 Hyracotherium

cating the power to which another number or expression is to be raised). Exponential functions involve the constant *e* (2.71828). Exponential growth is an increase in population size by a constant factor for each unit of time. It applies, for example, to a population that doubles in a short period of time. A graph of the population against time produces a curve that is very nearly flat at the beginning but then shoots almost directly upward. *See* **J-shaped curve**.

extinction

The complete disappearance of a species. Historic extinctions are believed to have occurred because species were unable to adapt quickly enough to a naturally changing environment. Mass extinctions are episodes during which whole groups of species have become extinct. The best known mass extinction is that of the dinosaurs, other large reptiles and various marine invertebrates about 65 million years ago.

The current wave of mass extinctions is largely due to human destruction of habitats, as in the tropical forests and coral reefs; it is far more serious and damaging than the mass extinctions of the past because of the speed at which it is occuring – in some places, 100 to 1000 times faster than the natural rate. Artificial climate changes and pollution (*see* **global warming**) also make it less likely that the biosphere can recover and evolve new species to suit a changed environment. The current rate of extinction is difficult to estimate, because most losses occur in the tropical rainforest, where the total number of existing species is not known; but it is estimated to be one species per day. *See also* **endangered species**.

fallow

Describing land left for a season after being ploughed and harrowed in order to allow it to recuperate. Fallow land is used in some modern **crop rotation** sequences and is particularly common in farming systems in the developing world (for example, shifting cultivation) which do not have access to fertilizers to maintain soil fertility.

famine

A severe shortage of food that affects a large number of people. Famine may be the result of drought, singly or in combination with population pressure or warfare.

fecundity

The rate at which an organism reproduces, as distinct from its ability to reproduce (fertility). In vertebrates, fecundity is usually measured as the number of offspring produced by a female each year.

feedback

The control or modification of a system by its effects or results. For example, a change in the status of the **trophic level** of one species affects all levels that exist above it in the **food chain**.

fern

Any plant of the class Filicales, related to horsetails and clubmosses. Ferns are spore-bearing (not flowering) plants, and most are perennial, spreading by creeping stems (rhizomes). The leaves, known as fronds, vary widely in size and shape. Most species are landliving, but a few species live in water. Bracken, a widespread fern, is often a troublesome weed on acid grassland because it is not eaten by most sheep and rabbits, and it recovers very quickly after burning.

fertilizer

An **agrochemical** containing some or all of a range of about 20 chemical elements necessary for healthy plant growth. Fertilizers are used to rectify the deficiencies of poor or depleted soil. They may be either organic (natural) or inorganic (artificial). Externally applied fertilizers may be in excess of plant requirements, and tend to drain away to adversely affect lakes and rivers (*see* **eutrophication**). In view of environmental concerns, research has now focused on the modification of crop plants themselves so as to allow nitrogen to be assimilated directly from the atmosphere.

field capacity

The maximum quantity of water that can be held in the porous spaces of a soil (by capillary action) after any excess has drained away. Also called field moisture capacity.

filter-feeder

An aquatic organism that ingests large quantities of plankton through the use of filters which trap the microscopic organisms in the plankton. Whales are typical filter-feeders.

fish

An ectothermic aquatic vertebrate that uses gills for obtaining oxygen from water and has a two-chambered heart There are three main groups: bony fish (for example, codfish and tuna), cartilaginous fish (sharks and rays) and jawless fish (hagfish and lampreys). Bony fish constitute the majority of living fish (about 20,000 species).

fisheries

Fishing grounds. They can be classified by type of water (freshwater or marine); catch (such as salmon or tuna fishing); or fishing method (diving, stunning or poisoning, har-

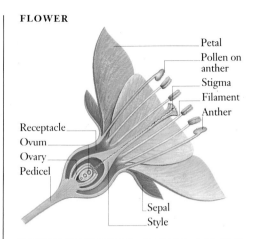

FLOWER

Petal
Pollen on anther
Stigma
Filament
Anther
Receptacle
Ovum
Ovary
Pedicel
Sepal
Style

pooning, trawling or drifting). Most of the world's catch comes from the oceans. The main production area is the upper pelagic zone, the relatively thin surface layer of water that can be penetrated by light, allowing photosynthesis by phytoplankton. There is also a large demand for freshwater fish, such as salmon, trout and carp. These inhabit ponds, lakes, rivers or swamps, and some species have been successfully cultivated (fish farming). In recent years, overfishing of certain fishing grounds has led to serious depletion of stocks and heated confrontations between countries using the same fishing grounds.

flower

The reproductive unit of a **flowering plant** (angiosperm), typically consisting of four whorls of modified leaves: sepals, petals, stamens and carpels. These are borne on a central axis or receptacle. Variations in their size, color, number and arrangement of parts are related to the method of pollination and are characteristic of a particular species. Flowers that are adapted for wind pollination have reduced or absent petals and sepals and long, feathery stigmas that hang outside the flower to trap airborne pollen; whereas the petals of flowers pollinated by insects are usually brightly colored.

flowering plant

A general term for a plant (angiosperm) that bear flowers. Sometimes the term includes both conifers and related species (gymnosperms) as well as angiosperms, in which the cones of conifers and cycads are referred to as flowers. Usually, however, the angiosperms and gymnosperms are referred to collectively as seed plants (spermatophytes).

food chain

The hierarchical feeding relationships between organisms: each depends on the next lowest member of the hierarchy for its food.

Food is transferred from primary producers (**autotrophs**, the first trophic level), which are principally plants, to a series of consumers. These consumers (**heterotrophs**) include herbivores, carnivores and decomposers, which break down the dead bodies and waste products of all four groups (including their own), ready for recycling. The concept has been used by environmentalists to show how poisons and other forms of pollution can pass from one animal to another. In reality, organisms have varied diets, so the term food chain is an oversimplification. *See* **food web**.

CONNECTIONS

CYCLES, CHAINS AND WEBS **68**

ENERGY TRANSFORMATION **70**

PRIMARY PRODUCERS **72**

THE CONSUMERS **74**

PYRAMIDS AND WEBS **76**

AGROCHEMICALS AND THE BIOSPHERE **124**

---

food web
A complex sequence of interlinked **food chains** that shows the feeding relationships between organisms in a natural community.

fossil
A relic of an animal or plant preserved by natural processes in rocks. Fossils may be formed by refrigeration (for example, Arctic mammoths in ice), carbonization (leaves in coal), formation of a cast (dinosaur or human footprints in mud) or mold (the form of a shell when the shell itself has disappeared), or mineralization of bones, teeth or shells. The study of fossils is called paleontology.

fossil fuel
Any fuel, such as coal, oil and natural gas, derived from the fossilized remains of plants. Fossil fuels are a nonrenewable energy source. They also release carbon dioxide, thus causing its accumulation in the atmosphere. Extraction of coal and oil can cause environmental pollution, and burning coal contributes to problems of **acid rain** and the **greenhouse effect**. *See* **alternative energy** and **global warming**.

CONNECTIONS

THE CARBON CYCLE **82**

DISRUPTING THE CARBON CYCLE **84**

ENDLESS ENERGY **88**

---

frost
A weather condition that occurs when the temperature of the air falls below freezing.

Water in the atmosphere condenses and freezes, is deposited as ice crystals on the ground (ground frost) or on objects (hoar frost). Because cold air is heavier than warm air, ground frost is more common.

fruit
The ripened ovary of a flowering plant that contains the seed or seeds. Fruits protect the seeds during their development and aid in their dispersal. Fruits that are dispersed by animals are often edible, sweet, juicy and colorful. When eaten they provide vitamins, minerals and enzymes, but little protein. When fruits are eaten by animals, the seeds are passed out with the feces.

Fruits are divided into agricultural categories on the basis of the climate in which they grow. They can also be divided botanically into those that are dry and those that become fleshy, or according to which flowers parts are incorporated. The structure of a fruit consists of the pericarp or fruit wall, which may be divided into a number of distinct layers. Sometimes parts other than the ovary (such as the receptacle) are incorporated into the fruit structure, resulting in a false fruit or pseudocarp, such as the apple and strawberry. Fruits may be dehiscent (opening to shed their seeds) or indehiscent (remaining unopened and being dispersed as a single unit).

fungicide
A chemical used to control fungal diseases in plants and animals. Inorganic and organic compounds containing sulfur are widely used. *See* **pesticide**.

Gaia hypothesis
The idea embraced by New Age thinkers and writers that the Earth's living and non-living systems form an inseparable whole that is regulated and kept adapted for life by the activities of living organisms themselves. The planet therefore functions as a single organism, like a giant cell. According to the hypothesis, because life and the environment are so closely linked, there is a need for humans to understand and maintain the physical environment and living things around them: the well-being of the planet matters more than that of any individual species. The Gaia hypothesis was elaborated by the British scientist James Lovelock in the 1970s. *See also* **holistic**.

CONNECTIONS

A LIVING PLANET **48**

THE GLOBAL CLIMATE **50**

PRESERVING DIVERSITY **142**

FRUIT

gasohol
A fuel for the internal combustion engine of cars, consisting of gasoline with the addition of 10 to 20 percent alcohol. The alcohol is made by destructive distillation (heating in the absence of air) of wood – that is, the process is based on a renewable resource – rather than on nonrenewable **fossil fuels**.

gene
The basic physical unit of inheritance that controls a particular characteristic of an organism. A gene is encoded by a specific segment of DNA whose nucleotide bases are aligned in a particular sequence. The term gene refers to the inherited factor that consistently affects the production of particular protein in an individual. The gene occurs at an exact point (locus) on a specific chromosome and may have several variants (**alleles**), each specifying a particular form of that character; for example, the alleles for red or white snapdragon flowers. Some alleles show **dominance**: they mask the effect of other alleles (known as **recessives**) when both are present. Genes undergo mutation and recombination to produce the variation of individuals on which the process of **natural selection** operates.

CONNECTIONS

BIOLOGICAL ALTERNATIVES **126**

PRESERVING DIVERSITY **142**

---

gene pool
The total sum of **alleles** (variants of genes) possessed by all the living members of a given population or species. Populations of organisms that have been reduced to low levels may encounter a number of problems as a result of a diminished gene pool; for example, lack of ability to respond to **natural selection** because of a lack of genetic diversity, reduction in breeding ability due

to increased biological relatedness (*see* **inbreeding**) and the increased expression of harmful alleles leading to the reduced fitness of offspring (for example, the appearance of hereditary diseases).

genetic engineering

The popular term for the deliberate manipulation of genetic material by biochemical techniques such as recombinant **DNA** technology, in which DNA from different **chromosomes** is combined. This can be for pure research, or to breed functionally specific plants, animals or bacteria. In genetic engineering, the splicing and reconciliation of genes is used to increase knowledge of cell function and reproduction, but it can also have practical applications. For example, plants grown for food could be given the ability to fix nitrogen, found in some bacteria, and so reduce the need for fertilizers. *See also* **biotechnology**.

genetic drift

See **evolutionary biology**.

genetic resource

An environment, such as a rainforest, that is rich in genetic diversity. *See* **biodiversity**.

geothermal energy

Energy derived from natural steam, hot water or hot rocks in the Earth's crust. Water is pumped down through an injection well where it passes through joints in the hot rocks. It rises to the surface through a recovery well and may be converted to steam or run through a heat exchanger. Dry steam may be used to drive turboalternators to produce electricity.

germination

The first stages of growth in a seed, spore or pollen grain. It occurs when seeds are exposed to favorable conditions of moisture, light and temperature, and when any factors causing **dormancy** have been removed. The process begins with the intake of water. In seeds, the embryonic root (radicle) is normally the first organ to emerge, followed by the embryonic shoot (plumule). Germination ends when the first true leaves emerge.

global warming

The projected climate change attributed to excessive carbon dioxide in the **atmosphere**, which acts like a thermal blanket, preventing the escape of heat. The United Nations Environment Program estimates that by the year 2025 average world temperatures will have risen by 1.5°C, with a consequent rise of 20 centimeters in sea level caused by melting of the polar ice caps. Low-lying areas, in-

GENETIC ENGINEERING

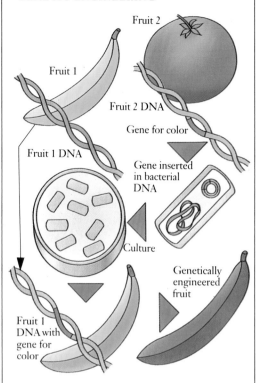

Fruit 2

Fruit 1

Fruit 2 DNA

Gene for color

Fruit 1 DNA

Gene inserted in bacterial DNA

Culture

Genetically engineered fruit

Fruit 1 DNA with gene for color

cluding some entire countries, would be threatened by flooding, and the distribution of crops would be affected by the change in climate. However, predictions about global warming are not yet accurate; they rely on incomplete computer modeling.

greenhouse effect

The increase in the temperature of the Earth's atmosphere caused by solar radiation, absorbed by and re-emitted from the Earth's surface, being prevented from escaping by various gases in the air. The main greenhouse gases are carbon dioxide (from the burning of fossil fuels and from forest fires), methane (mostly from agriculture) and **chlorofluorocarbons** (CFCs). Water vapor is another greenhouse gas. The effect was named by the Swedish scientist Svante Arrhenius, but it was predicted in 1827 by the French mathematician Joseph Fourier. *See* **global warming**.

Green Revolution

Beginning in the 1940s, the gradual development and spread of high-yield cereals to help feed the rapidly growing populations of lesser-developed countries such as Mexico, India, Pakistan and the Philippines. The term was coined in 1968. It was widely used following the award of the 1970 Nobel Peace Prize to the United States agronomist Norman Borlaug for breeding crop varieties to improve the ratio of food output to popu-

lation growth. A related package of measures included controlling water supplies and increasing the use of **fertilizers** and chemical protection. Highly successful at first, the increase in output achieved by the Green Revolution leveled off by the early 1970s, and technological and socioeconomic problems began to emerge: dependence on imported technology (often based on fossil fuels); greater susceptibility to pests; increased pollution of water supplies; widening gaps in income and regional inequalities. Green Revolution countries are heavy users of insecticides, which often contain toxic chemicals banned in the West. The growing surplus of grain worldwide has also unbalanced the international market, with important economic implications.

groundwater

Water that occupies spaces in rocks and soil, usually derived from rainwater that has soaked into the ground or, more rarely, from aquifers and other subterranean sources.

CONNECTIONS

NATURAL CYCLES **78**

WATER FACTORS **94**

Gulf Stream

A strong, warm current that flows across the Atlantic Ocean from the Gulf of Mexico to western Europe. Its path, 80 kilometers wide, follows the eastern seaboard of the United States up to Newfoundland before it veers across the Atlantic toward the British Isles. The northern section of the Gulf Stream is called the North Atlantic Drift. It keeps the climate along the coasts of northwestern Europe mild in winter and prevents ocean ports from freezing over.

habitat

The local environment in which a particular organism lives. The diversity of habitats within the **ecosystem** is enormous. Many can be considered inorganic or physical: the Arctic ice cap, a cave or a cliff face. Others are more complex: a woodland or a forest floor. Some habitats are so precise that they are called microhabitats; an example is an area under a stone where a particular type of insect lives. *See also* **biome**.

habitat loss

The more or less permanent destruction of a **habitat**, usually as a result of the activities of humans, such as **deforestation** of tropical rainforests and the removal of vegetation that leads to **desertification**. One consequence of habitat loss is the drastic reduction in the numbers of one or more species that may eventually result in their extinction.

halosere

The stages in a primary plant **succession** beginning under **saline** conditions. Also known as halarch succession. *See also* **sere**.

herbicide

A chemical used to destroy plants or check their growth, also known as a weedkiller. Selective herbicides are effective with cereals because they kill broadleaf plants without affecting plants with grasslike leaves. Their widespread use has lead to a dramatic increase in crop yields. However, it can also lead to polluted soil and water supplies and to the killing of birds and smaller animals.

herbivore

Any animal that feeds on green plants (including shoots, roots, seeds, fruits and nectar) or **algae**. Herbivores are more numerous than other types of animals, and as **primary consumers** they form a vital link in the food chain between plants (producers) and carnivores (higher level consumers). *See also* **carnivore** and **omnivore**.

heterotrophic

Describing an organism that is a consumer – one that takes in complex **organic compounds** as food because it cannot synthesize them itself from simple inorganic substances. This large group – the heterotrophs – includes bacteria, protozoa, fungi and all animals. The main exceptions are plants, which are **autotrophic** (producers).

CONNECTIONS

ENERGY TRANSFORMATION 70
PRIMARY PRODUCERS 72
THE CONSUMERS 74

---

hibernation

A state of greatly reduced metabolism and suspended bodily activity by which some animals (for example, bats, hedgehogs and bears) survive the winter months of food scarcity and cold weather. *See also* **estivation**.

holistic

The belief that systems may be understood properly only when viewed in their entirety. The **Gaia hypothesis** is one application of this belief to ecological systems.

holozoic

Describing an organism that feeds on relatively complex pieces of organic matter; for example, most animals.

HOMEOSTASIS

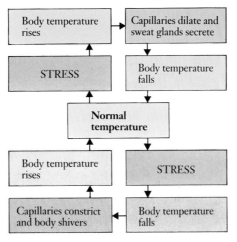

homeostasis

The maintenance of a constant internal state within a biological system. It may be considered either in terms of interactions between organisms in a community. Or it may refer to the internal environment of an individual organism, where stable levels of pH, salt concentration, temperature and blood sugar are important for the efficient functioning of the enzyme reactions within the cells.

hormone

A chemical secretion produced by an endocrine gland which helps to control body functions. The main glands are the thyroid, parathyroid, pituitary, adrenal, pancreas, thymus, ovary and testis. Hormones bring about changes in the functions of various organs according to the body's requirements. The pituitary gland, at the base of the brain, is a center coordinating hormone secretion.

Humboldt Current

Also called the Peru Current, a cold ocean current that flows north from the Antarctic along the western coast of South America as far as the Equator and then veers westward. It has the effect of cooling winds that blow from the Pacific Ocean onto the coast, keeping the western slopes of the Andes dry.

humidity

The moisture content of air or another gas. Relative humidity is the amount of water vapor in air divided by the total amount it can hold (at a particular temperature).

humus

The decomposed or partly decomposed organic component of soil, dark in color. It has a higher carbon content than the original material and a lower nitrogen content, and is an important source of minerals in soil fertility. Soil near the surface is rich in humus.

hunter–gatherer

A practitioner of an early form of agriculture in which people hunted prey for food and foraged for fruits and vegetables a certain area before progressing to a new location. This method – analogous to crop rotation – was gradually replaced as livestock became domesticated and crop plants were cultivated on permanent plots.

hurricane

A revolving storm in tropical regions (also known as a tropical **cyclone** or typhoon). Hurricanes originate at latitudes 5° to 20° north or south of the Equator when the surface temperature of the ocean is above 27°C. A central calm area, called the eye, is surrounded by inwardly spiraling winds (anticlockwise in the Northern Hemisphere, clockwise in the Southern) of up to 320 kmh. They are accompanied by lightning and torrential rain, and can cause extensive damage.

hydroelectricity

Electricity generated using the power of flowing water, usually water in a fast-flowing river or water that has been stored behind a dam. The water rotates turbines which, in turn, drive alternators to generate electricity. Such power generation is "clean" because it does not involve the burning of fossil fuels, but damming rivers can have disastrous effects on the local ecosystem.

hydrology

The study of the chemistry and physics of inland water, both frozen and liquid, above and below ground. It is applied to major civil engineering projects such as irrigation schemes, dams and hydroelectric power, and in planning water supply.

hydroponics

A method of cultivating plants without soil. It allows increased planting density and a year-round growing season. The plants are grown in peat or gravel, through which a solution of artificial nutrients is circulated. Hydroponics is best suited to small-scale greenhouse cultivation and is often used in arid climates or in crowded countries with little land to spare for agriculture.

hydrosere

The stages in a primary plant **succession** that starts in aquatic habitats, such as lakes, and progresses to drier conditions. It is also known as hydrarch succession. *See also* **sere**.

iceberg

A mass of ice floating on the sea, about 80 percent of which is submerged. Icebergs are formed from glaciers or ice barriers that

KEYWORDS

reach the sea in the polar regions, and then break off and drift toward temperate latitudes, sometimes endangering shipping.

immigration
The movement of individuals to an environment that is not their native one, in order to settle there. Immigration is most common in mobile species such as mammals, birds, fish and insects. *See* **population dynamics**.

immunization
Conferring immunity to infectious diseases by artificial methods. The most widely used technique is vaccination, in which fragments of a modified infectious agent are administered to stimulate the production of antibodies, thus producing active immunity.

inbreeding
Also known as endogamy, the mating of closely related individuals. The degree depends on the relationship between the parents. It is undesirable because it increases the risk that offspring will inherit copies of rare deleterious recessive **alleles from both parents, leading to** disabilities. The resulting decrease in viability or fertility is known as inbreeding depression. *See also* **outbreeding**.

indicator species
A plant or animal whose presence or absence in an area indicates certain environmental conditions, such as soil type, or low levels of dissolved oxygen in rivers. Many plants prefer either alkaline or acid soil, and certain trees are found only in soils with high concentrations of aluminum. Some lichens are sensitive to sulfur dioxide in the air, and their absence indicates **atmospheric pollution**.

CONNECTIONS
ATMOSPHERIC POLLUTION **132**
WATER POLLUTION **134**

Industrial Revolution
The period of history, beginning in the late 18th century, that saw the rise of the use of machines in industry. It began with the application of the steam engine and the development of railways, and shifted the balance of power from landowners and their farms (*see* **Agricultural Revolution**) to capitalists in industry, marking the beginning of serious and cumulative manmade damage to the Earth's ecological systems.

inorganic
Describing material that does not have the structure or characteristics of living organisms. Inorganic chemicals are compounds that do not contain carbon and that are not manufactured by organisms. However, carbon dioxide and other simple oxides and sulfides are also considered to be inorganic compounds, although they contain carbon. *See also* **organic**.

insecticide
Any **pesticide** used to kill insects. Among the most effective insecticides are synthetic organic chemicals (for example, **DDT**) which are chlorinated hydrocarbons. These chemicals, however, remain in the environment and are poisonous (*see* **accumulation**). Insecticides prepared from plants, such as derris and pyrethrum, are safer but need to be applied frequently and carefully.

integrated control
A system of pest management that uses the principles of **biological control** together with general pesticides and other practices such as crop rotation. *See* **organic farming**.

intensive farming
Maximizing crop yields and livestock produce, using efficient machinery, "factory farming" techniques, agricultural chemicals and genetically modified organisms. It can lead to pollution from fertilizers, deterioration of soil structure, and the loss of habitats (for example, through deforestation).

intercropping
The practice of increasing the productivity of agricultural land by growing additional crops in the periods between the growing seasons of the main crops, or by growing a secondary crop in the spaces between rows of the main crop plants.

intertidal
Relating to the zone of the sea shore exposed between the high and low water marks.

INTERCROPPING

Cabbages and carrots   Corn (maize) and beans

An intertidal **community** is characterized by **zonation** of the organisms that live there.

invasion
The encroachment or intrusion of animals or plants onto an area to which they are not native, especially onto newly cleared or recently burned land. *See* **succession**.

inversion layer
An uncommon condition in which the layer of air next to the Earth is cooler than the overlying layer. It usually occurs in valleys or basins, and may give rise to **smog** over cities.

invertebrate
An animal without a backbone. Invertebrates make up over 95 percent of animal species and include sponges, cnidarians, flatworms, nematodes, annelid worms, arthropods, mollusks, echinoderms and primitive aquatic chordates, such as sea squirts and lancelets. *See also* **vertebrate**.

ionosphere
A layer in the Earth's upper atmosphere, above 75 kilometers in altitude, which absorbs high-energy solar radiation.

CONNECTIONS
THE GLOBAL CLIMATE **50**
DISRUPTING THE CARBON CYCLE **84**
ATMOSPHERIC POLLUTION **132**

irrigation
The artificial supply of water to agricultural areas by dams and channels in order to grow crops. Irrigation tends to concentrate salts, ultimately causing soil infertility (*see* **salinization**). Rich river silt is retained at dams, leading to the impoverishment of the land and fisheries farther downstream. An example is the channeling of the annual Nile flood in Egypt, which has been carried out from earliest times and is now controlled by the controversial Aswan High Dam.

irruption
The sudden emigration of a population (partial or total), usually in response to overcrowding or food shortage. The mass emigrations of lemmings are an example.

isolation
The separation of populations that allows **divergent evolution**. Isolated populations often show many differences in their **phenotypes** (observable characteristics) from other populations of the same species. This is the case, for example, for many species found on islands. *See also* **evolutionary biology**.

J-shaped curve

The shape of the graph of a given population plotted against time, when the population grows exponentially and limiting factors (*see* **environmental resistance)** are absent. *See also* **S-shaped curve.**

jet stream

A 300-mile-wide channel of wind with a velocity of more than 150 km/h. It occurs at altitudes between 10 and 16 kilometers in the upper troposphere or lower stratosphere and moves horizontally from west to east in both Hemispheres. Jet streams usually occur at temperate latitudes. A sharp temperature difference exists across the air streams, with warmer air lying to the right of the jet stream in the Northern Hemisphere and to the left in the Southern Hemisphere.

K-species

Species adapted by natural selection to living in stable situations where competition may be very intense. *K*-species are less able to take advantage of particular opportunities to expand than are *r*-species. They tend to be **density dependent,** have a relatively long lifespan of a year or more, and have relatively low reproduction rates.

larva

The stage between hatching and adulthood in species in which the young and adults have a different appearance and way of life. Examples include tadpoles, maggots and caterpillars. Larvae are more common among invertebrates, some of which have two or more larval stages. Among vertebrates, only amphibians and some fish have a larval stage. The process by which larvae change into another stage, such as a pupa (chrysalis) or adult, is **metamorphosis.**

latitude

A series of imaginary lines that run in a circle around the Earth parallel to the Equator (which is 0° latitude), used to locate points on the surface. Latitude is measured up to 90° north and south of the Equator, with latitudes of 90° found at the poles. The circles are widest at the Equator and become smaller as they range north or south.

leaching

The removal of the soluble constituents of rock, ore and soil by the action of water. Fertilizers may leach out of the soil and drain into rivers, lakes and ponds, causing **eutrophication** and water pollution. In tropical areas, leaching of the soil after the destruction of forests removes scarce nutrients and can lead to a dramatic loss of soil fertility. The leaching of soluble minerals in soils can cause the formation of distinct soil horizons as different minerals are re-deposited at successively lower levels.

leaf

A flattened lateral outgrowth on the stem of a plant that is the primary organ of **photosynthesis** in most plants. Most leaves are composed of a sheath (leaf base), a petiole (stalk) and a lamina (blade), which has a network of veins that carry water and nutrients to the leaf and the products of photosynthesis from the leaf to the rest of the plant. The structure is made up of mesophyll cells surrounded by the epidermis and, usually, a waxy cuticle, which prevents excessive transpiration (water loss). The epidermis contains small pores (stomata), through which gas exchange occurs. A simple leaf is undivided; a compound leaf has several leaflets.

legume

Any plant of the family Papilionacae, which has a pod containing dry seeds that are released by lengthwise splitting of the pod down both sides. The family includes peas, beans, lentils, clover and alfalfa (lucerne). Legumes are important because of their specialized roots, which have nodules that contain bacteria which fix nitrogen from the air and increase the fertility of the soil. *See also* **nitrogen cycle** and **nitrogen fixation.**

lethal dose

The minimum dose of a pesticide that is fatal to the target species. It increases as a pest builds up **resistance** to a particular pesticide.

light

Electromagnetic radiation in the visible range, having a wavelength from about 400 nanometers in the extreme violet to about 770 nanometers in the extreme red. Light energy is used in the production of carbohydrates in **photosynthesis** and can be considered the principal energy source for virtually all living organisms. Light is an important **abiotic** factor in the environment.

CONNECTIONS

CYCLES, CHAINS AND WEBS **68**

ENERGY TRANSFORMATION **70**

ENDLESS ENERGY **88**

limiting factor

Any environmental factor that restricts the distribution or activity of an organism. Levels of light or of carbon dioxide may be limiting factors for plants because they are both necessary for photosynthesis and the production of carbohydrates. Photosynthesis slows down and eventually stop as the levels of light decrease.

longitude

An imaginary line that joins the Earth's poles. The line that passes through London, England (called the prime or Greenwich meridian) is designated 0° longitude. All other locations have longitudes of up to 180° east or west of this meridian, forming a series of imaginary rings. **Latitude** and longitude specify the position of any place on Earth.

malnutrition

Ill health resulting from a lack of sufficiently nutritious food, as opposed to starvation, which results from lack of any food at all.

mammal

Any member of a class of vertebrates called Mammalia, which contains about 4000 species. Mammals are characterized by having mammary glands in the female; these produce milk, which is used for suckling the young. Other mammalian features include hair (very reduced in some species, such as whales), a middle ear formed of three small bones (ossicles), a lower jaw consisting of two bones only, seven vertebrae in the neck, a thorax separated from the abdomen by a diaphragm, a four-chambered heart, and no nucleus in the red blood cells. Mammals are divided into three groups: placental mammals, in which the young develop inside the uterus (womb), receiving nourishment from the blood of the mother via the placenta; marsupials, in which the young are born at an early stage of development and develop further in a pouch (marsupium) on the mother's body; and monotremes, in which the young hatch from an egg outside the mother's body. The monotremes are the least evolved and have been largely displaced by marsupials and placentals, so that there

LEAF

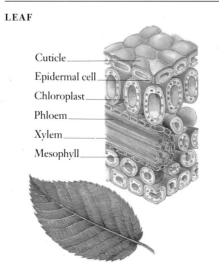

Cuticle

Epidermal cell

Chloroplast

Phloem

Xylem

Mesophyll

are only a few types surviving (platypus and echidnas). Placentals are the most widespread mammals.

mangrove

A tree or shrub, especially of the mangrove family Rhizophoraceae, found in the muddy swamps of tropical and subtropical coasts and in salt marshes and estuaries. By sending down aerial roots from their branches, mangroves rapidly form close-growing thickets. Their timber is waterproof and resists marine worms. Mangrove swamps are rich breeding grounds for fish and shellfish, but are rapidly disappearing in many countries.

marginal land

Poor-quality agricultural land that is likely to yield a low return. It is the last land to be brought into production and the first land to be abandoned. Examples are desert fringes in Africa and mountain areas in North America and Europe.

microbiome

An ecological community of organisms that occupies a relatively small area, such as a woodland, lake or even a hedgerow. *See also* **biome** and **habitat**.

> **CONNECTIONS**
>
> THE WORLD'S BIOMES **58**
> HABITATS AND NICHES **64**
> SUCCESSION **100**

microclimate

The climate of a very local area. Significant differences can exist between the climates of two neighbouring areas; for example, a town is usually warmer than the surrounding countryside (forming a heat island), and a woodland cooler, darker and less windy than an adjacent area of open land. Microclimates play a significant role in agriculture and horticulture, because different crops require different growing conditions.

midrib

The main vein running through a **leaf.**

migration

The regular movement of animals, chiefly birds, marine mammals and fish, to distant breeding or feeding grounds. It may be seasonal or part of a single lifecycle. Species migration is the spread of a species' home range over many years. A journey that takes an animal outside its normal home range is called individual migration; if the animal does not return it is called removal migration. *See* **immigration** and **emigration.**

mineral

Any natural inorganic substance with a particular chemical composition and a regular internal structure. Minerals are the constituents of rocks, and include commercially used materials such as ores and fossil fuels.

monoculture

A system of agriculture in which only one crop is grown year after year. In developing countries it is often a cash crop, grown on large plantations. Cereal crops in the industrialized world are also frequently grown on a monoculture basis. Monoculture allows production to be tailored toward one crop, but it is a high-risk strategy because the crop may fail (due to pests, disease or bad weather) or world prices for the crop may fall. Also, monoculture without **crop rotation** results in reduced soil fertility, contributes to soil erosion and encourages the spread of specific pests.

> **CONNECTIONS**
>
> THE GREEN REVOLUTION **118**
> FEEDING THE WORLD **120**
> OVEREXPLOITATION **128**

monohybrid inheritance

The pattern of inheritance shown by plants or animals that have different versions of the gene for a particular trait. The basic principle (and pattern) is shown by a monohybrid cross: breeding two parents that are pure-breeding for different forms (alleles) of the gene. This gene may code for a feature such as green or yellow seed color. The offspring of these parents, known as the F_1 (first filial) generation, are all hybrid (heterozygous) with respect to this gene, containing one allele for each color. When this happens, one **allele** is usually dominant (D) over the other (d) and only its color is expressed – so the F1 seeds are all the same color as those of the parent containing the dominant allele. When the F1 offspring breed among themselves, three combinations of these alleles may be found in their offspring, the F_2 generation: DD, Dd and dd. DD and Dd show the dominant color, dd the recessive color. In an infinitely large population, these colors would appear in the F2 generation in the ratio 3 to 1 (DD/Dd to dd). This is called the monohybrid ratio. The Austrian biologist Gregor Mendel first carried out experiments of this type, crossing artificially bred plants such as peas. The same mechanism underlies all inheritance, but in most plants and animals, so many genetic differences produce external appearance (phenotype) that such simple patterns are not evident.

monsoon

A seasonal wind pattern that brings heavy rain to southern Asia. It blows from the northeast (toward the sea) in winter and rainladen from the southwest (toward the land) in summer. The monsoon cycle is believed to have started about 12 million years ago with the uplift of the Himalayas.

moss

A class of plants within the division Bryophyta that are found in damp conditions and possess leafy stems giving rise to leafless stalks bearing capsules with distinct lids. Spores formed in the capsules are released and grow to produce new plants.

mudflat

A tract of low muddy land composed of fine silts found near estuarine regions that are covered at high tide and exposed at low tide. The silts tend to retain fine organic matter. Few species are adapted to live on mudflats; however, those that are may reach large numbers because of the high productivity of the area. Typical inhabitants include clams, crabs, annelid worms and gastropod mollusks, which in turn support large numbers of wading birds.

mutation

A change in the genetic constitution of an organism produced by an alteration in its **DNA** (the material that makes up the hereditary material of all living organisms). Mutations, which are the raw material of evolution, result from mistakes during replication (copying) of DNA molecules. Only a few of these mistakes improve the organism's performance and are therefore favored by **natural selection**. Mutation rates are increased by certain chemicals and by radiation. Common mutations include the omission or insertion of a base (one of the chemical subunits of DNA); these are known as point mutations. Larger scale mutations include removal of a whole segment of DNA or its inversion within the DNA strand. Not all mutations affect the organism, because there is a certain amount of redundancy in the genetic information.

natural resources

Resources that occur naturally; for example, coal and wood. They may be **renewable** or **nonrenewable**.

> **CONNECTIONS**
>
> ENDLESS ENERGY **88**
> OVEREXPLOITATION **128**
> CONSERVING AND RESTORING **140**

natural selection

The process by which gene frequencies in a population change through certain individuals producing more descendants than others because they are better suited to their environment. Because most environments are slowly but constantly changing, natural selection enhances the reproductive success of members of the population that possess favorable characteristics. The process is slow, relying on random variation in the genes of an organism due to mutation and on genetic recombination during sexual reproduction. It is the main driving process of evolution. *See* **evolutionary biology**.

niche

The position or status of a species in its **ecosystem**, including its nutrition, feeding times, temperature, moisture, and the way it reproduces. It is believed that no two species can occupy exactly the same niche, because they would be in direct competition for resources at every stage of their life cycle.

nitrate

A salt of nitric acid. Nitrates are widely used in the agrochemical and pharmaceutical industries. They are the most water-soluble salts known and play a major part in the **nitrogen cycle**. Nitrates in the soil, whether naturally occurring or from inorganic or organic fertilizers, are absorbed by the roots of plants and are their chief source of nitrogen. *See* **nitrogen cycle** and **nitrogen fixation**.

nitrogen cycle

The sequence of chemical reactions by which nitrogen circulates through the ecosystem. Inorganic nitrogen compounds (such as nitrates) in the soil are absorbed by plants and turned into organic compounds (such as proteins). Some of this nitrogen is eaten by herbivores and passed on to carnivores which feed on the herbivores. The nitrogen is ultimately returned to the soil as excreta and when organisms die, and converted back to inorganic form by decomposers (bacteria and fungi) before being taken in by plants again. Some nitrogen is cycled from plant to soil to atmosphere and back again through the activities of soil-dwelling bacteria. *See also* **denitrification**.

nitrogen fixation

The process by which atmospheric nitrogen is converted into nitrogen compounds by microorganisms, such as blue–green bacteria (cyanobacteria). There are two kinds of nitrogen-fixing bacteria in soil: one is free-living and the other lives symbiotically in the root nodules of legumes. The process indirectly makes nitrogen available to plants.

Nitrogen fixation also occurs in the ocean, where free-living bacteria and blue-green bacteria are the fixing agents. Several chemical processes duplicate nitrogen fixation to produce fertilizers.

nitrogen oxides

Compounds such as nitric oxide (NO) and nitrogen dioxide (NO_2), produced by the burning of **fossil fuels** and their derivatives, which can form a major part of atmospheric pollution that leads to **acid rain**.

nocturnal

Activity associated with the hours of darkness, as opposed to daylight. *See also* **diurnal**.

nomadism

A system of livestock farming in which animals are taken to fresh pastures in different locations. Large herds may lead to overgrazing and **desertification**. Nomadism is under threat due to increasing enclosure of land.

nuclear power

Energy derived from the controlled nuclear fission of uranium or plutonium in a reactor. Heat from the reaction is used to generate steam to power turbines to drive electricity generators. Opponents of nuclear power highlight the problems of decommissioning old power stations and safely disposing of accumulated nuclear waste.

nutrition

The means by which an organism obtains the chemicals it needs to live, grow and reproduce. In **autotrophic** nutrition, the Sun's energy is used in photosynthesis. **Heterotrophic** nutrition involves ingesting, digesting and absorbing food. This food directly or indirectly comes from plant matter; therefore, all animals ultimately depend on plants for their nutrition. *See* **food chain**.

ocean

Any great mass of salt water. There are five oceans on Earth – the Antarctic, Arctic, Atlantic, Indian, and Pacific. Together they cover almost 70 percent (363,000,000 square kilometers) of the total surface area of the Earth. Oceans are complex biomes (communities) and play a major role in the water cycle, as well as shaping the climates of different regions. Underneath the water are a continental shelf, the gently sloping sea floor at the edge of land; a continental slope, which extends out from the continental

OCEAN

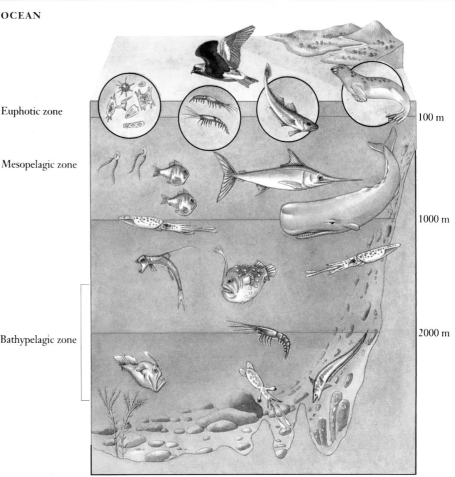

Euphotic zone — 100 m

Mesopelagic zone — 1000 m

Bathypelagic zone — 2000 m

shelf; and beyond it the deep-sea or abyssal plain, where the ocean is deepest – an average depth of 4000 meters below sea level.

High temperatures of 35°C have been recorded in shallow tropical waters, and even the -1.9°C measured during the Antarctic winter is much warmer than the temperature on land near the South Pole. Salinity varies depending on the temperature of the water, the amount of precipitation and evaporation, and the movement of currents. Because water warms up and cools down more slowly than land, oceans and ocean currents modify the climate of nearby landmasses.

oligotrophic
Describing bodies of water that are poor in plant nutrients. They are unproductive and often have very clear water due to the sparse growth of planktonic organisms. The term can also be applied to any lake in which the lowest level of cold water does not become depleted of oxygen during the summer months. *See also* **eutrophic**.

omnivore
An animal that feeds on both plant and animal material.

open-cast mining
The extraction of minerals, particularly metal ores and coal, by excavating them from deposits at or near the surface. This method does not require shafts or tunnels, but it causes extensive scarring of the land.

organic
The general term for material that has the structure or characteristics of living organisms. Organic chemicals contain carbon and can be manufactured by living organisms. *See also* **inorganic**.

organic farming
Farming without agrochemicals. Compost, manure, seaweed or naturally-derived substances are used instead. Growing a crop of a nitrogen-fixing plant (legume), then ploughing it back into the soil, is another method (*see* **crop rotation**). Weeds can be controlled by hoeing, mulching (covering with manure, straw or black plastic), or burning. Organic farming produces food with minimal residues and reduces pollution. It is labor intensive and therefore more expensive, but uses less fossil fuel.

CONNECTIONS

THE NITROGEN CYCLE **86**

AGROCHEMICALS AND THE BIOSPHERE **124**

BIOLOGICAL ALTERNATIVES **126**

---

outbreeding
Sexual reproduction between individuals not closely related. Crossing between two genetically distinct lines may lead to hybrid vigor – increased fitness due to genetic diversity. It is also known as outcrossing or exogamy. *See also* **inbreeding**.

overcrowding
Very high population densities in areas without sufficient resources, caused by unrestrained multiplication of a species. *See* **carrying capacity**.

overexploitation
Exhausting natural resources by excessively heavy use – for example, by excess cultivation or fishing. *See* **intensive farming**, **overfishing** and **whaling**.

overfishing
The catching of fish at rates that exceed the capacity of the stock to replenish itself by reproduction, resulting in a decline in population. For example, in the North Atlantic, herring have been fished to the verge of extinction and the codfish and haddock populations are severely depleted. In the developing world, use of fleets of large factory ships, often from industrialized countries, has depleted stocks for local people who cannot obtain protein from any other source.

overgrazing
The result of poor farming practice in which animals are allowed to consume completely the local supply of plants on which they feed. Possible consequences of overgrazing include soil **erosion** and **desertification** once the land's plant cover has been removed. *See also* **nomadism**.

oxygen
A colorless, odorless, tasteless, nonmetallic, gaseous element. Oxygen is a byproduct of photosynthesis and the basis for **respiration** in plants and animals. As the basis of water (H_2O), it is also of crucial importance to maintaining life on Earth.

ozone
A highly reactive pale-blue gas made up of three atoms of oxygen. It is formed when a molecule of the stable form of oxygen (O_2) is split by ultraviolet radiation or electrical discharge. It forms a thin layer in the upper atmosphere, which protects the Earth from damaging ultraviolet radiation. There has been recent concern about holes appearing in the ozone layer over the polar regions, thought to be caused by the release of **chlorofluorocarbons** (CFCs). An increase in the amount of ultraviolet radiation reach-

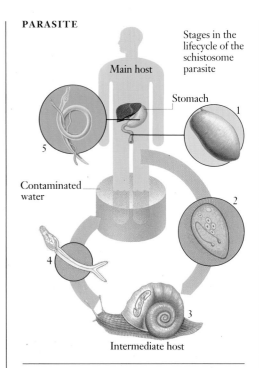

PARASITE

Stages in the lifecycle of the schistosome parasite

Main host

Stomach

Contaminated water

Intermediate host

ing the Earth could cause changes in climate, increased skin cancer, and damage to crops, vegetation, livestock and wildlife both on land and in the ocean. At lower atmospheric levels ozone is a pollutant and contributes to the **greenhouse effect**. Near the ground, ozone can cause asthma attacks, stunted plants and corrosion of certain materials. It is produced by the action of sunlight on air pollutants (*see* **smog**).

paleontology
The study of past life forms, encompassing the structure of ancient organisms and their environment, evolution and ecology, as revealed by their **fossils**. The practical aspects of paleontology are based on using the presence of different fossils to date particular rock strata (layers laid down throughout time) and to identify rocks that were laid down under particular conditions, for instance giving rise to the formation of oil.

parasite
An organism that lives on (ectoparasite) or in (endoparasite) another organism (the host) and depends on it for nutrition. Some parasites cause relatively little damage to the host, whereas others give rise to characteristic diseases. Obligate parasites can survive and reproduce only as parasites. Facultative parasites can also live as saprophytes – that is, by absorbing dead organic matter.

pedology
The study of the distribution and morphology of **soil**. Pedology is an important aspect of agricultural science.

pelagic

The upper layers of water in the **ocean** – down to about 100 meters – where most marine life is found. Plenty of sunlight is able to penetrate the water at those depths, thus allowing **photosynthesis** in plants, and attracting the animals in the local marine **food chain**. The pelagic zone is also known as the euphotic zone. *See also* **benthic**.

perennial

Any plant that continues to live from year to year. Herbaceous perennials have aerial stems and leaves that die each autumn. They survive the winter by means of an underground storage (perennating) organ, such as a bulb or rhizome. Trees and shrubs or other woody perennials have stems that persist above ground throughout the year, and may be either deciduous or evergreen.

permafrost

Ground that is permanently frozen. It covers approximately 26 percent of the world's land surface in alpine areas and near glacial zones, and gives rise to a poorly drained form of grassland known as **tundra**. At less extreme latitudes, the upper layers of permafrost – some 4 meters deep – may thaw during summer, when they are particularly vulnerable to erosion by human activity in the area.

persistent

Describing pesticides that are resistant to decay in the environment. They may accumulate in the environment; in particular they may be concentrated through food chains, so although the primary producers and herbivores may contain only low concentrations, the top level carnivores may contain substantial amounts of these chemicals. Persistent pesticides may also give rise to resistant strains of pests (*see* **resistance**).

pest

Any insect, plant, fungus, rodent or other living organism that humans consider harmful, other than those that directly cause human diseases. Most pests damage crops or livestock; others damage buildings and other structures and destroy food stores, and can also carry disease.

pesticide

Any chemical agent used to combat pests. There are three main types: insecticides (to kill insects), fungicides (to combat fungal diseases, molds and so on) and herbicides (to kill plants, mainly those considered weeds). Pesticides can cause a number of pollution problems through spray drifting onto surrounding areas, direct contamination of users or the public and as residues on food.

Potent synthetic products, such as DDT and dieldrin, are highly toxic to wildlife and often to human beings. Safer pesticides are based on organic phosphorus compounds, but they still present potential hazards to health. *See* **persistent**.

CONNECTIONS

INTENSIVE FARMING 122
THE EFFECTS OF AGROCHEMICALS 124

phenotype

The characteristics and appearance of an organism. *See also* **genotype**.

phloem

Vascular tissue found in many plants. It transports sugars and other food materials from the leaves to all other parts of the plant. Phloem is usually found in association with **xylem**, the water-conducting tissue, but unlike the latter it is a living tissue.

photoperiodism

The biological mechanism that determines the timing of certain activities by responding to changes in day length. The flowering of many plants is initiated in this way, regulated by a light-sensitive pigment, phytochrome. The breeding seasons of many temperate-zone animals are also triggered by increasing or decreasing day length. Autumn-flowering plants (for example, chrysanthemum and soy bean) and autumn-breeding mammals (such as goats and deer) respond to nights that are shorter than a certain length; spring-flowering and spring-breeding ones (such as lettuce and birds) are triggered by longer days.

photosynthesis

The process by which green plants and photosynthetic algae trap light energy and use it to drive a series of chemical reactions, leading to the formation of carbohydrates. The plant must possess **chlorophyll** and have a supply of carbon dioxide and water. The chemical reactions of photosynthesis occur in two stages. During the light reaction, sunlight is used to split water into molecular oxygen, hydrogen ions and electrons, with oxygen given off as a byproduct. In the dark reaction, for which sunlight is not required, the hydrogen ions and electrons are used to convert carbon dioxide into carbohydrates. Photosynthesis depends on the ability of chlorophyll to capture the energy of sunlight. Other pigments, such as carotenoids, are also involved in capturing light energy and passing it on to chlorophyll. **Heterotrophic** organisms depend on the supply of organic material produced by chlorophyll-containing plants during photosynthesis, so they also are dependent on sunlight as their ultimate source of energy.

CONNECTIONS

CYCLES, CHAINS AND WEBS 68
ENERGY TRANSFORMATION 70
PRIMARY PRODUCERS 72
PYRAMIDS AND WEBS 76
THE CARBON CYCLE 82

phytoplankton

The plant component of **plankton**. It consists mainly of algae, such as diatoms and desmids, and they carry out almost all of the **photosynthesis** in the oceans. All animal life in the open sea depends ultimately on phytoplankton, which is the basis of food chains that lead to fish, whales and sea birds.

pioneer species

Species that are the first to colonize and thrive in new areas. Coal tips, recently cleared woodland and new roadsides are areas where pioneer species quickly appear. As the habitat matures other species take over (*see* **colonization** and **succession**). Pioneer species are usually **r-species**.

plagioclimax

An artifical **climax** community that forms following the clearing of existing vegetation. Examples are heathland and downland.

plankton

Small, often microscopic, forms of photosynthetic organisms (phytoplankton) and animals (zooplankton) that live in the upper layers of fresh and salt water and are an important food source for larger animals. Plankton float or swim very feebly and are moved by winds, waves and currents. Marine plankton is concentrated in areas where rising currents bring mineral salts to the surface. *See* **phytoplankton** and **zooplankton**.

podsol (or podzol)

A type of soil associated with plants that favor acid conditions, such as heather and pine. It consists of an upper layer of **humus** overlying a bleached layer, on top of a hard "pan" containing deposits of metals such as aluminum and iron (leached from the layer above). It is found in regions of high rainfall and low evaporation.

polar

Describing the climate typical of the areas around the North and South Poles (within the Arctic and Antarctic Circles). Polar re-

POLLUTION

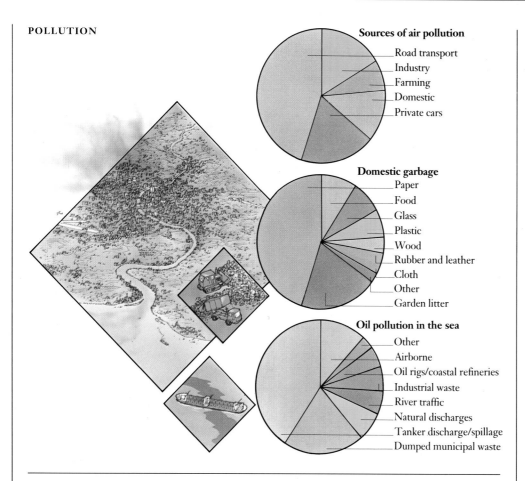

Sources of air pollution
- Road transport
- Industry
- Farming
- Domestic
- Private cars

Domestic garbage
- Paper
- Food
- Glass
- Plastic
- Wood
- Rubber and leather
- Cloth
- Other
- Garden litter

Oil pollution in the sea
- Other
- Airborne
- Oil rigs/coastal refineries
- Industrial waste
- River traffic
- Natural discharges
- Tanker discharge/spillage
- Dumped municipal waste

gions are characterized by low termperatures and the absence of trees (because of the unfavorable conditions such as the shortness of the growing season).

pollen

Microspores of seed-producing plants that eventually give rise to the male gametes. In flowering plants (angiosperms), pollen is produced within anthers; in most cone-bearing plants (gymnosperms) it is produced spore sacs or male cones. A pollen grain is typically yellow and, when mature, has a hard outer wall. Pollen of insect-pollinated plants is often sticky, spiny and larger than the smooth light grains produced by wind-pollinated species.

pollination

The process by which pollen is transferred between plants. The pollen grains are transferred to the female stigma in flowering plants (angiosperms) and to the female cone in cone-bearing plants (gymnosperms). Fertilization takes place after the growth of the pollen tube to the ovary and formation and release of the male gametes. Self-pollination occurs when pollen is transferred to a stigma of the same flower or to another flower on the same plant; cross-pollination occurs when pollen is transferred to another

plant by external pollen-carrying agents, such as wind, water, insects or birds.

pollution

The hazardous effects on the environment caused by byproducts of human activity, principally industrial and agricultural processes (noise, smoke, automobile emissions; chemical and radioactive effluents in air, seas and rivers; pesticides, radiation, sewage) and household waste. Pollutants may enter the **food chain** and be passed on from one organism to another. They are frequently harmful and may cause additional damaging developments, such as **acid rain**. Much recent concern has centered on the fact that chemical pollution often travels great distances and may affect large areas or even the entire planet (causing climatic change). *See* **global warming** and **greenhouse effect**.

CONNECTIONS

AGROCHEMICALS AND THE BIOSPHERE **124**

ATMOSPHERIC POLLUTION **132**

WATER POLLUTION **134**

population

A group of organisms of the same species that live together in one geographical area.

Within a population, all individuals capable of reproduction have the opportunity to reproduce with other mature members. The size of the population is determined by the **carrying capacity** of the **habitat**.

population dynamics

The study of changes in population densities (that is, changes in the numbers of individuals in a specified area). Areas may be compared by looking at variations in population density and at population cycles (fluctuations in the size of a population). Such cycles are often caused by **density-dependent** mortality: high mortality due to overcrowding causes a sudden decline in the population, which then gradually builds up again. A population "explosion" is the swift and usually unpredictable increase in population size, which may be due to natural events such as a change in climate, or to human intervention, as when new species are introduced to an area. A population "crash" is a sudden decline brought about by the inability of a habitat to sustain its population. It tends to occur after a population has exceeded its **carrying capacity** and consequently has exhausted basic resources such as oxygen, food and living space. Population crashes are associated with algae, hares and lemmings. Population cycles may result from the interaction of predator and prey.

CONNECTIONS

POPULATION STUDIES **108**

POPULATION CURVES **110**

OUT OF CONTROL **116**

precipitation

The condensation of water from the atmosphere, which falls to the Earth. It includes rain, snow, sleet, hail, dew and frost.

POPULATION

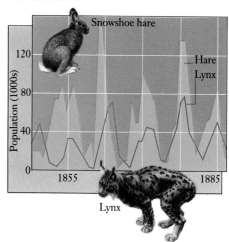

predator

Any animal that kills another animal (the prey) for food and, therefore, is a secondary (or tertiary) consumer in the **food chain**.

prey

Any animal hunted or captured by a **predator** for food.

primary forest species

Any forest species (nonpioneer) that can establish and grow in very small gaps (for example, the space left by the death of a single tree). Secondary forest species (pioneers) require large cleared areas.

primary producer

Also called **autotroph**, an organism at the lowest level of a **food** chain which is ultimately responsible for feeding the **consumers (heterotrophs)** above it. Most primary producers are plants or algae. Their productivity – their gain in weight during a particular period of time – is generally expressed in kilograms per square meter per year, and is a measure of the input of energy into the **ecosystem**. *See also* **producer**.

CONNECTIONS

ENERGY TRANSFORMATION **70**
PRIMARY PRODUCERS **72**
THE CONSUMERS **74**
PYRAMIDS AND WEBS **76**

---

producer

An organism that does not need to take in organic compounds. Plants and photosynthetic algae are producers: they trap light energy through photosynthesis and produce food (carbohydrates) which they make available to consumers. *See* **autotrophic**, **food chain** and **primary producer**.

productivity

Primary productivity is the amount of organic material produced in a given time by the primary producers (**autotrophic** organisms) in a particular environment. It is expressed as mass (g/m^2) or energy (mJm^{-2}). Net productivity is the amount of organic material produced in excess of that used during its production (that is, in respiration) and represents the potential food for the consumers of an ecosystem. Secondary productivity is the iuncrease in **biomass** of the consumers and decomposers (the heterotrophs).

protein

Any of many complex, biologically important polymeric substances composed of amino acids joined by peptide bonds. Other types of bond, such as sulfur bonds, hydrogen bonds and cation bridges between acid sites, are responsible for creating a protein's characteristic three-dimensional structure. Proteins are essential to all living organisms. As enzymes, they regulate all aspects of metabolism. Structural proteins (such as keratin and collagen) make up skin, claws, bones, tendons and ligaments; muscle proteins produce movement; hemoglobin transports oxygen; and membrane proteins are responsible for regulating the movement of substances into and out of cells.

protozoa

A group of single-celled organisms without rigid cell walls. Some, such as amoeba, ingest other cells, but most are saprotrophs (*see* **saprozoic**) or **parasites**.

pyramid of biomass

A graphic representation of the biomass in each trophic level of a food chain. The difference in biomass between each level represents losses of organic material or energy in activity and metabolism, and in waste.

pyramid of numbers (biotic)

A graphic representation of the **trophic levels** within a stable **food chain**. As the number of individuals in higher trophic layers decreases, there are more primary producers at the base of the pyramid than primary consumers in the trophic layer above, then more secondary consumers than primary consumers, and so on. Animals at higher levels of the pyramid are usually larger in size.

r-species

Species that are adapted by natural selection to maximize their rate of increase and are thus able to rapidly colonize recently disturbed sites. They tend to be **density independent** and have a short life span. *See also* K species and **pioneer species**.

CONNECTIONS

ADAPTATION AND EVOLUTION **104**
POPULATION STUDIES **108**
COLONIZATION STRATEGIES **114**

---

radiation

The dissemination of radiant energy from a source, such as electromagnetic waves, acoustic waves and alpha and beta particles emitted in radioactivity. Radiation of heat is the transfer of heat by infrared rays.

radioactivity

The process that establishes an equilibrium in parts of the nuclei of unstable radioactive substances, ultimately to form a stable, non-radioactive element. This is most frequently accomplished by the emission of alpha particles (helium nuclei), beta particles (electrons and positrons) or gamma radiation (electromagnetic waves of very high frequency), which are all biologically harmful. The safe disposal of waste containing radioactive substances, produced in laboratories, hospitals and, principally, as a byproduct of nuclear power generation, is a major environmental problem. *See also* **nuclear energy**.

rain

Precipitation from clouds in which separate drops of water fall to the Earth's surface. The drops are formed by the accumulation of fine droplets that condense from water vapor in the air. Condensation is usually brought about by rising and subsequent cooling of air. Rain can form in three main ways: frontal (or cyclonic), orographic (or relief) and convectional. Frontal rainfall takes place at the boundary, or front, between a mass of warm air from the tropics and a mass of cold air from the Poles. Orographic rainfall occurs when an airstream is forced to rise over a mountain range. Convectional rainfall, associated with hot climates, is brought about by rising and abrupt cooling of air that has been warmed by the extreme heat of the ground surface. The water vapor carried by the air rapidly condenses and so rain falls heavily. *See* **climate**.

rainforest

Dense forest found in equatorial regions, where the climate is hot and wet. Heavy rainfall results as the moist air brought by the converging trade winds rises because of the heat. Rainforests comprise some of the most complex and diverse ecosystems on the planet and help to regulate global weather patterns. More than half of the world's tropical rainforest is in Central and South America; most of the rest is in Africa and southeast Asia. It provides most of the oxygen needed for plant and animal respiration. Tropical rainforests are characterized by a great diversity of species, usually of tall broadleaved evergreen trees, with many climbing vines and ferns, some of which are a main source of raw materials for medicines.

Tropical rainforest once covered 14 percent of the Earth's land surface, but it is now being destroyed at an increasing rate as valuable timber is harvested and the land cleared for agriculture, causing problems of **deforestation**. When deforestation occurs, the microclimate of the mature forest disappears; soil erosion and flooding become major problems because rainforests protect the shallow tropical soils. Clearing of the rainforests may lead to **global warming** of the atmosphere and contribute to the **greenhouse effect**.

reclamation
The conversion of swamp, marsh, desert or other wasteland into land suitable for cultivation or settlement.

recycling
The processing of industrial and household waste (such as paper, glass, and some metals and plastics) so that it can be reused, thus saving expenditure on scarce raw materials, slowing down the depletion of nonrenewable resources and helping to reduce pollution. The natural passage of elements such as carbon and nitrogen through the environment and their return to the beginning of the cycle (*see* **carbon cycle** and **nitrogen cycle**) may also be considered to be a form of natural recycling.

reef
A calcareous bank, made up of the accumulated skeletons of **coral**, which forms an important marine habitat. Fringing reefs build up on the shores of continents or islands, the living animals mainly occupying the outer edges of the reef. Barrier reefs are separated from the shore by saltwater lagoons, which may be as much as 30 kilometers wide; there are usually navigable passages through the barrier into the lagoon. Atolls resemble a ring surrounding a lagoon and do not enclose an island. They are usually formed by the gradual subsidence of an extinct volcano, the coral growing up from where the edge of the island once lay.

regeneration
The natural renewing of growth that follows destruction, for example, forest regrowth after felling. Such forest regeneration is essential for the long-term stability of many resource systems. *See also* **succession**.

renewable energy
Energy produced from any source that replenishes itself. Most renewable systems rely on solar energy directly or through the weather cycle as wave power, hydroelectric power or wind power, or solar energy collected by plants (alcohol fuels, for example). In addition, the gravitational force of the Moon and Sun can be harnessed through tidal power stations and the heat energy trapped in deep-lying rocks is harnessed through geothermal energy systems. *See also* **alternative energy**.

reproduction
The process by which a living organism produces other organisms similar to itself. Reproduction may be either asexual or sexual. Asexual reproduction produces identical individuals because there is only one parent and so no fusion of sex cells. Sexual reproduction involves two individuals and produces genetic variation as a result of fertilization of the female ovum (egg) by the male gamete (sperm or pollen). Species that reproduce only once in their life (for example, salmon and most bamboos) are said to be semelparous; those that reproduce on two or more occasions are said to be iteroparous.

reptile
Any member of the vertebrate class Reptilia, which includes snakes, turtles, alligators and crocodiles. Unlike amphibians, reptiles have eggs with hard or leathery shells and yolks which are laid on land and produce fully formed young. Breathing is by means of lungs. Some snakes and lizards retain their

Fringing reef
round volcano

Barrier reef
round island

Atoll

eggs and give birth to live young. Reptiles cannot control their body temperature and the skin is usually covered with scales. Their metabolism is slow, and in large snakes intervals between meals may be months. Reptiles date back more than 300 million years. Many forms are now extinct, including the orders Pterosauria, Plesiosauria, Ichthyosauria and Dinosauria.

resistance
The property of an organism that is not readily attacked by a parasite, disease or drug. Genetic resistance can be acquired by environmental selection pressures, as in the case of certain pests which become resistant to **pesticides** that persist in the environment or have been used over a long period of time.

resource
Any commodity that can be used to satisfy human needs. Resources can be categorized into human resources, such as labor, supplies and skills, and environmental resources. Environmental resources may be further divided into renewable resources, such as water and timber (that is, those that can be used repeatedly, given appropriate management), and nonrenewable resources, such as metal ores and fossil fuels (that is, those of which there is a fixed supply, which will eventually be exhausted).

resource partitioning
The sharing of a resource such as food or nesting sites by two or more species. For example, some desert plants partition the restricted local water supply by dividing up access to the water through the use of roots of different lengths by different species.

respiration
The biochemical process by which food molecules are broken down (oxidized) to release energy in the form of **ATP**. In the first stage (glycolysis), in the cytoplasm, glucose is broken down to pyruvate, a form of anaerobic respiration (it does not require oxygen). In the second stage (the Krebs cycle) the pyruvate is further broken down by a cyclic series of reactions to produce carbon dioxide and water. This is the main energy-producing stage of respiration and requires oxygen. Glycolysis and the Krebs cycle are common to all organisms that respire aerobically. The Krebs cycle occurs in the mitochondrion.

restoration
The practice of returning a habitat to its original state; for example, replanting hedgerows and woodland on land devastated by poor agricultural practice.

rhizome

An underground stem that grows horizontally. Plants with rhizomes, such as many species of grass, often spread by means of these creeping underground stems, which may also act as storage organs to aid survival in adverse seasons.

river

Any long watercourse that flows down a slope along a channel. It originates at a point called its source and enters a sea or lake at its mouth. Along its length it may be joined by smaller tributaries. A river and its tributaries are contained within a drainage basin. The communities found in rivers and their adjacent lakes may vary considerably as a result of the different conditions existing in each biome: the directional and rapid flow of water in a river and the relatively stationary water in a lake produce different biomes.

CONNECTIONS

AQUATIC BIOMES 62

WATER FACTORS 94

WATER POLLUTION 134

RNA

Ribonucleic acid, a **nucleic acid** present in several forms in all cells. It consists of nucleotides containing the sugar ribose, linked by the bases adenine, guanine, cytosine and uracil (*see also* **DNA**). Messenger RNA, transfer RNA and ribosomal RNA all play a part in the mechanisms by which DNA directs the synthesis of proteins within a cell.

rock pool

An intertidal habitat formed when a retreating tide leaves pools of water in rock indentations on the shoreline.

root

That part of a vascular plant, usually underground, whose primary functions are anchorage and the absorption of water and dissolved mineral salts. Roots usually grow downward and toward water (*see* **tropism**). Some plants, such as epiphytic orchids, which grow above ground, produce aerial roots that absorb moisture from the atmosphere. Others, such as ivy, have climbing roots arising from the stems, which attach the plant to trees and walls. Symbiotic associations occur between the roots of certain plants (such as clover) and various bacteria that fix nitrogen from the air (*see* **nitrogen fixation**).

rotation

See **crop rotation**.

ruderal

Describing plants that inhabit rubbish heaps, wasteland, old fields or waysides.

ruminant

An even-toed hoofed mammal that chews cud (ruminates). Ruminants typically have a four-chambered stomach and no upper incisor teeth. Symbiotic bacteria in the stomach help digest the plant fibers of their herbivorous diet. Ruminants include cattle, sheep, goats, deer, antelopes and giraffes.

runoff

Water from rain or snow that flows off the surface of the land and into streams and rivers. Runoff that washes fertilizers from the surrounding agricultural land is a cause of environmental pollution. *See* **erosion** and **eutrophication**.

saline

Describing any solution that contains a metal salt resembling common rock salt (sodium chloride). The salinity of water is the total amount of such dissolved material, and it is usually expressed in terms of kilograms of material per million kilograms of water – parts per million (ppm) of total dissolved solids. On the average, seawater has a salinity of 35,000 ppm, of which 30,000 ppm is sodium chloride. *See also* **salinization**.

salinization

The accumulation of sodium, magnesium and potassium salts in water or soil. It usually occurs in arid and semi-arid areas where the rate of precipitation is less than the rate of evaporation. Salinization can also be caused by poor **irrigation** techniques associated with farming. Plants cannot survive in such conditions, and the area becomes unfit for cultivation. Salinization is an important factor in **desertification**.

CONNECTIONS

WATER FACTORS 94

SOIL FACTORS 96

FEEDING THE WORLD 120

salt marsh

A wetland habitat with characteristic vegetation that is tolerant of saline soil (halophytic vegetation). Salt marshes develop around estuaries and on the sheltered side of sand and shingle spits. They usually have a network of creeks and drainage channels by which tidal waters enter and leave the marsh. Salt marshes are found at higher latitudes than **mangrove** swamps, which cannot tolerate even minor frosts.

sampling

The practice of randomly selecting a small number of examples and assuming that they are representative of the total number of examples when performing statistical surveys.

sand

Loose grains of rock, sized 0.0625–2.0 millimeters in diameter, consisting chiefly of quartz, but owing their varying color to the presence of other minerals. Sand may eventually consolidate to form sandstone, a type of sedimentary rock.

saprotroph

Describing any organism that feeds on dead or decaying animal or plant matter; for example, house fly larvae (maggots). Saprotrophs cannot make food for themselves, so they are considered to be a type of **heterotroph**.

season

A division of a year that has a characteristic climate. In temperate latitudes, four seasons are recognized: spring, summer, autumn and winter. Tropical regions have two seasons – the wet and the dry. Monsoon areas around the Indian Ocean have three seasons: the cold, the hot and the rainy. The change in seasons is mainly due to the change in attitude of the Earth's axis in relation to the Sun and hence the position of the Sun in the sky at a particular place. Seasonal differences are more marked inland than near the coast, where the sea moderates temperatures.

secondary succession

A type of **succession** that starts in areas that have previously supported plant growth and which still retain some of the effects of this previous colonization (for example, remnants of soil and other organic matter, and even some seeds and plants that have survived after being burned or covered in flood silt). The changes on abandoned cultivated land are also secondary successions.

seed

The reproductive structure of higher plants (angiosperms and gymnosperms) derived from a fertilized ovule, and consisting of an embryo and a food store, surrounded and protected by an outer seed coat, called the testa, which is derived from the parent plant. Food for the seed is contained either in specialized nutritive tissue (the endosperm) or in the cotyledons of the embryo itself. In angiosperms the seed is enclosed within a fruit, whereas in gymnosperms it is usually naked and unprotected, once shed from the female cone. Following germination the seed develops into a new plant.

seed bank

A store of seeds in the soil that may lie dormant for years. Often the seeds germinate only when the soil has been disturbed, bringing them into contact with an environmental trigger, such as a change in temperature, carbon dioxide, light or water. The term is also used for a store of seeds that is kept in the and laboratory used to maintain genetic resources and diversity.

selection pressure

Factors that affect the course of evolution through **natural selection**. They include population density, competition, food availability and geographical attributes. *See also* **isolation** and **natural selection**.

sere

A series of plant communities (**succession**) that develop in a particular habitat. A lithosere is a succession starting on the surface of bare rock. A hydrosere is a succession in shallow fresh water, beginning with planktonic vegetation and the growth of weeds and other aquatic plants, and ending with the development of swamp. A halosere is a succession beginning in saline conditions. A plagiosere is the sequence of communities that follows clearing of existing vegetation.

CONNECTIONS

HABITATS AND NICHES 64

SUCCESSION 100

ADAPTING TO THE ENVIRONMENT 102

---

sewage

Liquid waste from a community. Domestic sewage consists of waterborne waste products from housing. Industrial sewage is normally from mixed industrial and residential areas. Sewage is conveyed through sewers to sewage works, to undergo treatment before being discharged into rivers or the sea. Raw sewage, or sewage that has not been treated adequately, is a source of water pollution and can cause **eutrophication**.

shrub

A perennial woody plant that produces several separate stems, at or near ground level, rather than the single trunk of most trees. A shrub is usually smaller than a tree, but there is no clear distinction between large shrubs and small trees. In a mature woodland the shrub level is the habitat that exists between the canopy and the woodland floor.

silt

A sediment intermediate in coarseness between clay and sand; its grains have a diame-ter of 0.002–0.02 millimeters. Silt is usually deposited in rivers, so the term is often used generically to mean a river deposit, as in the silting up of a channel. *See also* **mudflat**.

slash-and-burn

A simple agricultural practice in which natural vegetation is cut and burned, and the clearing then farmed for a few years until the soil loses its fertility, whereupon farmers move on and leave the area to regrow. Although this pratice is possible with a small, widely dispersed population, it becomes unsustainable with more people and is often a form of **deforestation**.

sleet

The precipitation of a mixture of rain and snow or small ice particles.

smog

Fog that contains impurities (such as unburned carbon and sulfur dioxide) derived from domestic fires, industrial furnaces, certain power stations, and petrol and diesel engines. Smog can cause substantial illness and loss of life, particularly among those with respiratory diseases, as well as damage to wildlife. Photochemical smog results from the action of sunlight breaking down chemicals by photochemical reactions to form additional harmful pollutants.

CONNECTIONS

DISRUPTING THE CARBON CYCLE 84

ENDLESS ENERGY 121

ATMOSPHERIC POLLUTION 132

---

snow

Precipitation in the form of soft, white, crystalline flakes of ice caused by the condensation in air of excess water vapor below freezing point. Light reflecting in the crystals, which have a basic hexagonal geometry, gives snow its white appearance.

soil

A loose covering of broken rocky material and decaying organic matter overlying the Earth's surface. Various types of soil develop under different conditions: deep soils form in warm wet climates and in valleys; shallow soils form in cool dry areas and on slopes. The organic content of soil is widely variable, ranging from zero in some desert soils to almost 100 percent in peats. In addition to organic and mineral materials, the volume of soil is almost equally composed of pores which are occupied by the soil atmosphere (resembling the atmosphere above ground) and soil water. The nature of a soil greatly influences the type of agriculture in a particular region. Mature soil contains a rich community of bacteria, fungi and other microorganisms, which help break down the organic matter and release nutrients into the soil water, where they become available to plants. A period of about 10,000 years undisturbed is required to produce mature soil, but this soil can be eroded within 50 years if the plant cover is removed and the surface is exposed to natural weathering and other damage. Immature soil (regolith) contains no humus and cannot support plant or animal life. Soil also ages, and will become worn out (senile) after about a million years of maturity, even if it is not abused by overfarming or overgrazing.

solar energy

The radiant energy from the Sun. The thermonuclear energy output of the Sun is estimated at more than 200 billion times that of any nuclear reactor on Earth, but Earth receives only one fifty-millionth of this energy; if it were much more, the Earth would be too hot to sustain most plant and animal life. About 50 percent of solar energy is scattered or absorbed by the atmosphere. What does get through to the planet is almost entirely absorbed by the oceans and feeds the water cycle. The remainder, some 5 to 9 percent, is used by plants for photosynthesis.

This energy can also be harnessed and put to practical use. It can be collected by mirrors and used as a source of heat, or it may be converted into electricity by photocells and solar panels. Solar energy has the great advantage of being both clean and renewable. It has been adapted to heat domestic hot water, to charge batteries, and to provide onboard power for satellite equipment, though large-scale applications still need developing. Large concave mirrors can focus the Sun's rays to raise the temperature of water

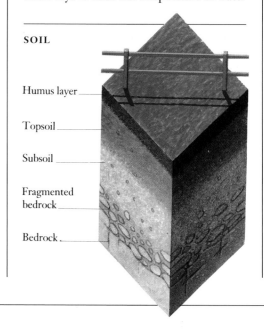

SOIL

Humus layer

Topsoil

Subsoil

Fragmented bedrock

Bedrock

to produce steam and generate electricity. The amount and intensity of solar radiation varies with geography and cloud cover – the British Isles receive only 2.5×10^8 cal/m^2 per year, compared with a world average of 15.3×10^8 calories.

CONNECTIONS

THE GLOBAL CLIMATE **50**
CYCLES, CHAINS AND WEBS **68**
ENDLESS ENERGY **88**

species

A distinguishable group of organisms that resemble each other or consist of a few distinctive types (polymorphs) and can interbreed to produce fertile offspring. Species are the lowest level in the system of biological classification. Related species are grouped together in a genus. In a scientific name, the first word indicates the genus and the second the species. Both names are written in italics, and the genus name starts with a capital letter: *Homo sapiens*. Within a species there are usually two or more separate populations, which may over a period of time become distinctive enough (through evolution) to be designated subspecies or varieties. These could eventually give rise to new species. Two closely related species may also interbreed to produce hybrid offspring, though these may not be fertile.

Around 1.4 million species have been identified so far, of which 750,000 are insects, 250,000 are plants and 41,000 are vertebrates. It is estimated that one species becomes extinct every day through **habitat** destruction.

A native species is one that has existed in a region from prehistoric times; a naturalized species is one known to have been introduced from another area by humans, but which now maintains itself; an exotic species is one that requires human intervention to survive. *See also* **evolutionary biology, K species, natural selection, pioneer species** and **r-species**.

S-shaped curve

The shape of the graph of the number of individuals in a population, plotted against time when the population grows and limiting factors are present. *See also* **J-shaped curve** and **population dynamics**.

CONNECTIONS

POPULATION STUDIES **108**
POPULATION CURVES **110**
CONTROLLING POPULATIONS **112**

stamen

The male reproductive organ of a flowering plant. The stamens are collectively referred to as the androecium. A typical stamen consists of a stalk with an anther, the pollen sac, at its tip. The number and position of the stamens are significant in the classification of flowering plants. Generally the more advanced plant families have fewer stamens, but they are often positioned more effectively so that the likelihood of successful pollination is not reduced. *See also* **flower,** flowering plant and **stigma.**

starch

A high molecular mass carbohydrate produced by plants as a food store. It consists of varying proportions of two glucose polymers (polysaccharides): amylose, a straight-chain polysaccharide, and amylopectin, a branched polysaccharide. The main dietary sources of starch are cereals, legumes and tubers, including potatoes.

stem

The main supporting axis of a plant which bears the leaves, buds and reproductive structures. It may be simple or branched, and usually grows above ground, although some grow underground, including **rhizomes** and corms. Stems contain a continuous vascular system that conducts water and food to and from all parts of the plant (*see* **phloem** and **xylem**). In plants exhibiting secondary growth, the stem may become woody, forming a main trunk (as in trees), or a number of branches from ground level (as in shrubs).

stigma

Part of the female reproductive organs of a flowering plant, consisting of the surface at the tip of a carpel which receives pollen. It often has short outgrowths, flaps or hairs to trap pollen, and may produce a sticky secretion to which the grains adhere. *See also* **stamen** and **flower.**

stomata

Pores in a plant's epidermis, usually on the underside of leaves. They are surrounded by a pair of guard cells which are crescent-shaped when the stoma is open but which can collapse to an oval shape, thus closing off the opening between them. Stomata control the exchange of gases between the internal tissues of the plant and the atmosphere for photosynthesis and respiration. They are also the main route by which water is lost from the plant (transpiration), and can be closed to conserve water, the movements being controlled by changes in turgidity of the guard cells.

stratosphere

The layer of the Earth's **atmosphere**, above the troposphere and the weather, that extends from an altitude of 12 kilometers up to about 50 kilometers. The upper stratosphere contains the **ozone layer,** which filters out harmful ultraviolet radiation from the Sun.

subsoil

Fragmented deposits that lie between the topsoil above it (which contains humus and nutrients for plants) and the bedrock below.

substrate

The surface to which an organism is attached or upon which it moves. The term is also used to refer specifically to a substance (such as agar) that provides the nutrients for the growth of microorganisms in the laboratory. In biochemistry, a substrate is a compound or mixture of compounds acted on by an **enzyme.**

succession

The progressive natural development that occurs in the vegetation in a given area from the time it is first colonized by plants (primary succession), or after it has been disturbed by fire, flood or clearing (secondary succession). A natural succession has two components: the physiologic, in which the organism responds to topographic features; and the **biotic**, in which organisms react with one another. If allowed to proceed undisturbed, succession leads naturally to a stable **climax community** (for example, oak and hickory forest or savanna grassland) which is determined by the climate and soil characteristics of the area.

CONNECTIONS

THE PHYSICAL ENVIRONMENT **92**
SUCCESSION **100**
ADAPTATION TO THE ENVIRONMENT **102**
ADAPTATION AND EVOLUTION **104**

sulfur dioxide

A poisonous choking gas. It is an oxide of sulfur formed during the burning of sulfur-containing **fossil fuels**, from which it becomes a component of **atmospheric pollution**. It reacts with oxygen in the atmosphere to form sulfur trioxide, which dissolves in water droplets to form **acid rain**.

surface water

Water that remains at or close to the land surface, as opposed to groundwater that occupies cracks in rocks and soil below the surface but above a layer of impermeable rock. *See* **water table**

SURVIVORSHIP CURVE

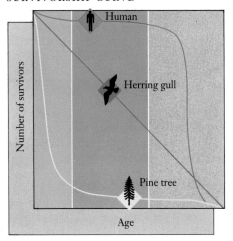

survivorship curve

A method of presenting demographic data for a particular population in a given environment. The curve is obtained by plotting the number or percentage of an original population surviving against time. Such a curve shows how the mortality and survival of individuals in the population is related to their age.

sustainability

The concept of using resources to improve life on Earth without reducing the Earth's ability to support life into the future. Through **sustainable** production and **conservation**, sustainability is a major goal of many environmental groups.

sustainable

Describing production or growth that is capable of being continued indefinitely, given the appropriate management. Sustainable differs from renewable. For example, wheat is renewable, because after the wheat has been harvested more wheat can be planted; however, it is not sustainable unless a high yield of wheat can be produced indefinitely year after year. If nutrients are lost from the soil or pests harmful to the wheat increase, wheat production will not be sustainable in the long term.

CONNECTIONS

FEEDING THE WORLD **120**

OVEREXPLOITATION **128**

CONSERVING AND RESTORING **140**

---

swamp

A wetland environment with soils that are waterlogged for most of the time. The dominant plant species are reeds and sedges. *See also* **bog** and **marsh**.

sweating

A method used to control body temperature in most mammals. Perspiration (sweat) is produced within specialized glands (sweat glands) in the skin and is secreted onto the surface. Evaporation of this surface perspiration then results in a cooling of the skin. Sweat also contains excreted waste products such as urea.

symbiosis

A close relationship between two organisms of different species in which both partners benefit from the association. A well-known example is the pollination of flowering plants by insects, in which the insects collect pollen while feeding on nectar and then carry it from one flower to another. Symbiosis is sometimes known as mutualism.

synecology

The study of the relationships within and between communities and their environment. *See also* **ecology** and **autecology**.

temperate

Describing a climate typical of mid-latitudes that is intermediate between the extremes of the polar and **tropical** climates. The temperate zone is considered to be the area of the Earth's surface that lies between the Arctic Circle and the Tropic of Cancer and the Antarctic Circle and the Tropic of Capricorn.

territory

A fixed area from which an animal or group of animals excludes other animals, usually of the same species. Territories may be held for many different reasons; for example, to provide a constant food supply, to monopolize potential mates or to ensure access to refuges or nest sites. The size of a territory depends in part on its function: some nesting and mating territories may be only a few square meters in extent, whereas feeding territories may be as large as hundreds of square kilometers.

thermal

A rising warm current of air (updraft) which may rise as high as 3 kilometers due to heating of the ground below, or due to turbulence produced when air masses collide with each other or with topographical obstacles such as mountains. Strong thermals occur during thunderstorms, early in the storm when warm air rises to the level where condensation and precipitation begin.

thermal depression

An area of low pressure that occurs when the intense heating of continental areas during the day leads to a convectional rise of air currents and decreased surface atmospheric pressure. Small thermal depressions are found over islands and peninsulas; larger ones in the Iberian peninsula and Arizona in the United States. Thermal depressions almost always occur in the summer.

thermal energy

Geothermal energy. It may derive from hot water and steam that originate in the Earth's interior, often giving rise to hot springs; or from drilling into hot igneous rocks just below the Earth's surface and injecting water to make steam. The water is a rich source of minerals; it can also be demineralized and used for domestic or industrial purposes. *See* **alternative energy** and **renewable energy**.

thermal erosion

Damage to **permafrost** areas that results from human activity. The construction of roads or buildings and laying of underground services requires the removal of surface plant cover, which provides insulation. This extends the active (uppermost) layer of the permafrost, which normally thaws during summer. With the thawing extending to an increased depth, the affected area quickly becomes a muddy bog, and the thawed soil is very susceptible to erosion by meltwater. Thermal erosion can be prevented by building roads and railways on insulating gravel beds and constructing buildings on piles to raise them above the permafrost.

thermocline

The layer of water in a lake that lies between the warmer uppermost layer (the epilimnion), which is regularly mixed by wind, and the cold noncirculating lower layer (the hypolimnion), where temperatre decreases steadily with depth. Within the thermocline the temperature decreases rapidly with depth. *See also* **inversion**.

THERMOCLINE

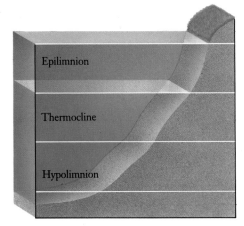

thermophile

A plant that can tolerate high temperatures.

tide

The rise and fall of sea level caused by the gravitational forces of the Moon and Sun. Dams and barrages can be used to harness the mechanical energy generated by tides for use in producing electricity. *See* **alternative energy** and **ocean**.

topography

A detailed description of the surface shape and composition of the landscape, comprising both natural and artificial features. Topographical features include the relief and contours of the land; the distribution of mountains, valleys and human settlements; and the patterns of rivers, roads and railways.

topsoil

The upper, cultivated layer of **soil**, which may vary in depth from 8 to 45 centimeters. It contains the decayed remains of vegetation (*see* **humus**), which plants need for active growth, along with a variety of soil organisms, including earthworms.

tornado

An extremely violent revolving storm with swirling, funnel-shaped clouds, caused by a rising column of warm air propelled by strong wind. A tornado can rise to a great height, but with a diameter of only a few hundred meters or less. Tornadoes move with wind speeds of 160–480 km/h, destroying everything in their path. They are common in the central United States and Australia. *See also* **dust storm**.

toxic

Describing a substance that is poisonous or harmful to living organisms. Radioactivity, air and water pollutants, and poisons ingested or inhaled (for example, lead from automobile exhausts, asbestos and chlorinated solvents) are some toxic substances that occur in the environment.

trade wind

The prevailing wind that blows toward the Equator from the northeast and southeast. Trade winds are caused by hot air rising at the Equator and the consequent movement of air from north and south to take its place. The winds are deflected toward the west because of the Earth's west-to-east rotation (*see* **Coriolis effect**). The unpredictable calms (the doldrums) lie at their convergence. The trade wind belts move north and south about 5° with the seasons.

transformation

A process by which genetic recombination occurs in microorganisms. During transformation, naked DNA from another organism or human agency passes directly into the cell. The term is also applied to the procedures in laboratory tissue culture which cause cells to grow and multiply out of control, rather like cancer cells.

transpiration

The loss of water vapor by evaporation through the leaves of a plant, mainly through the **stomata** but also to a lesser extent through the waxy cuticle. Transpiration results in a stream of water carrying dissolved nutrients being drawn up through the plant's transport system (*see* **xylem**).

tree

Any perennial plant with a woody stem, usually a single stem or trunk, made up of wood and protected by an outer layer of bark. There is no clear dividing line between trees and **shrubs**, but sometimes a minimum height of 2 meters is used to define a tree. A treelike form has evolved independently many times in different groups of plants. Among the angiosperms (flowering plants), most trees are dicotyledons. This group includes trees such as oak, beech, ash, chestnut, lime and maple, and they are often referred to as broadleaved trees because their leaves are broader than those of conifers, such as pine and spruce. In temperate regions angiosperm trees are mostly deciduous (that is, they lose their leaves in winter); in tropical regions most angiosperm trees are evergreen. There are fewer trees among the monocotyledons, although palms and bamboos belong to this group. The gymnosperms include many trees and they are classified into four orders: Cycadales (including cycads and sago palms), Coniferales (conifers), Ginkgoales (including only one living species, the ginkgo or maidenhair tree) and Taxales (including yews).

trench (oceanic)

A narrow, elongated depression at the ocean–continental margin. According to plate tectonics, trenches mark the site of destructive plate margins where an oceanic plate is forced down to the Earth's mantle beneath an adjacent plate. Ocean trenches are characterized by high levels of seismic activity. *See also* **ocean**.

trophic level

The position that is occupied by a species (or group of species) in a **food chain**. The main levels are primary producers (photosynthetic microorganisms, larger algae and plants), primary consumers (herbivores), secondary consumers (carnivores) and decomposers (bacteria and fungi).

tropical

Describing the climate typical of the tropics, an area lying between the Tropics of Cancer and Capricorn. These two points are the limits of the area of the Earth's surface in which the Sun can be directly overhead, and in which there is a mean monthly temperature of more than 20°C. Climates within the Tropics lie in parallel bands. Along the Equator is the intertropical convergence zone, characterized by high temperatures and year-round heavy rainfall (*see* **rainforest**). Along the Tropics themselves lie the tropical high-pressure zones, characterized by descending dry air and desert conditions. Between these, the conditions vary seasonally between wet and dry, producing the tropical grasslands.

Tropic of Cancer

The line of **latitude** 23.5° north of the Equator, which separates the northern temperate zone from the **tropics**.

Tropic of Capricorn

The line of **latitude** 23.5° south of the Equator, which separates the northern temperate zone from the tropics.

tropism

Directional growth of a plant, or part of a plant, in response to an external stimulus such as gravity or light. If the movement is directed toward the stimulus it is described as positive tropism; if away from it, it is negative tropism. Geotropism, for example – the response of plants to gravity – causes the root (positively geotropic) to grow downward and the stem (negatively geotropic) to grow upward. Phototropism occurs in response to light, hydrotropism to water, chemotropism to a chemical stimulus and thigmotropism (or haptotropism) to physical contact. Tropic movements result from a greater rate of growth on one side of the plant organ than the other.

KEYWORDS

tuber

A swollen underground stem, often used by a plant as a storage organ (or starch) by means of which it can survive the winter and from which it can then grow again. Potatoes are an example of tubers.

tundra

Treeless arctic and alpine regions of high latitude, resulting from the presence of **permafrost**. The vegetation consists mostly of grasses, sedges, heather, mosses and lichens. Tundra stretches in a continuous belt across northern North America and Eurasia. The term was originally applied to the topography of part of northern Russia, but is now used for all such regions.

typhoon

A violently revolving storm in the western Pacific Ocean (*see* **hurricane**).

ultraviolet (UV) radiation

Electromagnetic radiation from the Sun, in the wavelength range 400 to 10 nanometers. In large quantities it is harmful to animal life, but the filtering effect of the **ozone layer** means that only tolerable amounts reach the Earth's surface.

upwelling

The movment of deep water, usually off the coast of a continent, that brings nutrients from the bottom of the body of water to feed the organisms (such as plankton) that live nearer the surface. In oceans and large lakes, organisms die and sink to the bottom, where they decompose, providing food for detritus feeders. This is the basis of the **food chain.** However, this cycle occurs in deep water where there is little plant or animal life. Upwelling transports the nutrients to the surface, where they can participate in **photosynthesis**, and consequently enriches

UPWELLING

- Plankton
- Organisms die
- Upwelling
- Detritus feeders
- Organisms decompose

coastal areas, estuaries and continental shelf habitats, where plankton are found in high concentrations.

vaccination

Introducing any preparation of modified pathogens (viruses or bacteria) into the body, usually either orally or by a hypodermic syringe, to induce the specific antibody reaction that produces immunity against a particular disease.

variable

A quantitative characteristic of an individual (for example, the dimensions or weight of an organism). The term also describes any quantity that can have more than one value.

variation

The difference between individuals of the same species at corresponding stages of its life cycle. Typically, variation in size, behavior, biochemistry or coloring may be found. The cause of the variation is genetic, environmental or more usually a combination of the two. The origins of variation can be traced to the recombination of the genetic material during the formation of the gametes and, more rarely, to mutation. *See* **evolutionary biology**.

vegetation

The plants of a specific area considered as a whole, rather than taxonomically. Vegetation provides a significant amount of moisture in the atmosphere as a result of transpiration. Also, as a result of the latent heat absorbed by transpiration, areas with vegetation remain cool compared with areas of bare rock, sand or dry earth.

vertebrate

Any animal with a backbone. The 41,000 or so species of vertebrates include mammals, birds, reptiles, amphibians and fish. They include most of the larger animals, but in terms of numbers of species are only a tiny proportion of the world's animals. The zoological taxonomic group Vertebrata is a subgroup of the phylum Chordata. *See also* **invertebrate**.

water

A colorless, tasteless, odorless liquid. It is an oxide of hydrogen (H_2O) and can exist in three different physical states: solid (ice), liquid and gas (water vapor or steam). Water freezes at 0°C and boils at 100°C. When liquid, it is virtually incompressible; as it freezes, it expands by an eleventh of its volume. At 4°C, one cubic centimeter of water has a mass of one gram; this is its maximum density, forming the unit of specific gravity.

This property causes ice to form only at the surface of bodies of water, allowing aquatic life to survive below. Water has the highest known specific heat and is an efficient solvent, particularly when heated. Most of the world's water is in the sea; less than 0.01 percent is fresh water. Water covers more than 70 percent of the Earth's surface and occurs as standing water (oceans and lakes), running water (rivers and streams), ice (glaciers and polar ice caps), rain and vapor.

water cycle

The natural circulatory system by which water is cycled through the environment. Essentially it involves precipitation (rain or snow) falling on the land, from which water runs off into streams and rivers that eventually reach the oceans. **Ocean** water evaporates from the surface, and in the atmosphere water condenses to form droplets that fall as rain. Some water is "consumed" by plants and animals which, through transpiration or respiration, release it directly into the atmosphere as water vapor.

CONNECTIONS

---

watershed

A divide separating two drainage basins (catchment areas), usually a ridge of high ground.

water table

The level beneath which the ground is permanently saturated, thus forming the upper surface of the groundwater (water collected underground in porous rocks). The water table rises and falls in response to rainfall and the rate at which water is extracted. In many irrigated areas the water table is falling as a result of the extraction of water. Regions with high water tables and dense industrialization have problems with pollution of the water table. In the United States, New Jersey, Florida and Louisiana have water tables contaminated by both industrial wastes and saline seepage from the ocean. *See* **surface water**.

wave

A ridge or swell formed by wind or other causes in large bodies of water. The power of a wave is determined by the strength of the wind and the distance of open water over which the wind blows (the fetch). If waves strike a beach at an angle, the beach material is gradually moved along the shore (longshore drift), causing a deposition of material in some areas and erosion in others. Suitable devices can be used to harness the mechanical energy generated by waves for use in producing electricity (*see* **alternative energy**).

weather

The variation of climatic and atmospheric conditions at any one place, or the state of these conditions at a place at any one time. Such conditions include humidity, precipitation, temperature, cloud cover, visibility and wind. A region's climate is derived from the average weather conditions over a long period of time. The study of the weather is known as meteorology. *See* **climate**.

weathering

The process by which exposed rocks are broken down by rain, frost, wind and other elements of the weather. It differs from **erosion** in that no movement or transportation of the broken-down material takes place. Weathering can be physical (or mechanical), chemical or, more usually, a combination of both. Physical weathering includes such effects as the splitting of rocks by the alternate freezing and thawing of water trapped in cracks and exfoliation (flaking caused by the alternate expansion and contraction of rocks in response to extreme changes in temperature). Chemical weathering is brought about by a chemical change in the rocks affected. The most common form is caused by rainwater that has absorbed carbon dioxide from the atmosphere and formed weak carbonic acid. The acid reacts with certain minerals in the rocks and breaks them down.

weed

A plant growing in the wrong place, such as among crops. It is difficult to define what a weed is, because nothing distinguishes it from other plants except that, from a human viewpoint, it is in the wrong place.

whaling

The hunting and killing of whales for whale oil, used for food and cosmetics; for the large reserve of oil in the head of the sperm whale, used in the leather industry; and for ambergris, a waxlike substance from the intestines, used in making perfumes. There are synthetic substitutes for all these products. Whales are also killed for their meat, which is eaten by the Japanese and was used as pet food in the United States and Europe. The International Whaling Commission (IWC), established in 1946, failed to enforce quotas on whale killing until world concern about the possible extinction of the whale mounted in the 1970s. By the end of the 1980s, 90 percent of blue, fin, humpback and sperm whales had been wiped out. Low reproduction rates mean that even protected whale species are slow to recover.

wild

The general term given to a plant or animal not cultivated or domesticated by humans.

wilting

The loss of stiffness (turgor) in plants, caused by a decreasing wall pressure within the cells making up the supportive tissues because of a lack of water. Wilting is most obvious in plants that have little or no wood.

wind

The horizontal movement of air across the surface of the Earth. Wind is caused by the movement of air from areas of high atmospheric pressure (anticyclones) to areas of low pressure (cyclones, or depressions). Its speed is measured using an anemometer or by studying its effects on, for example trees, by using the Beaufort scale. On a global scale several major wind patterns can be identified (*see* **trade winds** and **monsoon**); however, these are modified locally by land and water. The direction of the wind is the direction from which it blows. *See also* **hurricane**, **cyclone** and **jet stream**.

woodland

Vegetation dominated by trees that form a distinct, but sometimes open, canopy; generally smaller than a forest. Temperate climates tend to support a mixed woodland habitat, with some conifers but mostly broadleaved and deciduous trees. In the Mediterranean region and parts of the Southern Hemisphere, most of the trees are evergreen. Temperate woodlands grow in the zone between the cold coniferous forest and the tropical forests of the hotter climates near the Equator. They develop in areas where the closeness of the sea keeps the climate mild and moist. Old woodland can rival tropical rainforest in the number of species it supports, but most of the animal species are hidden in the soil. A mature woodland is an example of a **climax community**.

CONNECTIONS

TERRESTRIAL BIOMES 60

SUCCESSION 100

DISAPPEARING FORESTS 136

xerophyte

A plant adapted to live in dry conditions. Common adaptations to reduce the rate of transpiration (and thus water loss) include smaller leaves, dense hairs over the leaf to trap a layer of moist air, water storage cells, sunken stomata and permanently rolled leaves or leaves that roll up in dry weather. Many desert cacti are xerophytes.

xerosere

The stages in a primary plant **succession** starting on dry ground (for example, the rocky areas left after the melting of the ice following an ice age). *See* **sere**.

xylem

Vascular plant tissue used to conduct water and dissolved mineral nutrients from the roots to other parts of a plant. It is composed of different types of cell, and may include long, thin, usually dead cells known as tracheids and conducting vessels; fibers; and thin-walled parenchyma cells . In most flowering plants, water is moved through these vessels. Non-woody plants contain only primary xylem, derived from the procambium. In trees and shrubs this is replaced for the most part by secondary xylem, formed by growth from the actively dividing vascular cambium, which allows increased transport as the plant increases in size. The cell walls of the secondary xylem are thickened by a deposit of lignin, providing mechanical support to the plant. *See also* **phloem**.

zonation

The occurrence in an area of distinct bands of vegetation, each with its own characteristic dominant species. Zonation can be seen on the seashore and on mountainsides. *See also* **succession**.

CONNECTIONS

THE WORLD'S BIOMES 00

SUCCESSION 100

zone

A region of the environment characterized by some distinctive feature; for example, the tropical zone or temperate zone, which are characterized mainly by temperature.

zoo

Zoological gardens, a place where animals are kept in captivity. Originally created purely for visitor entertainment and education, zoos have become major centers for the breeding of endangered species of animals.

zooplankton

The animal component of **plankton**, chiefly protozoans, small crustaceans (such as krill) and the larval stages of mollusks and other invertebrates.

A LIVING *Planet*

ALL LIVING THINGS are affected by their environment – the surroundings in which they live. Ecology is the study of the way in which each living thing interacts with its environment, both influencing and being influenced by it. Ecologists use the term ecosystem to describe this interdependent relationship, which includes both living (biotic) and nonliving (abiotic) factors. Their goal is to understand how ecosystems work.

Some ecologists support the Gaia hypothesis, which supposes that the Earth can be considered as a single super-organism. In this theory, every part of the planet, including the rocks, oceans, atmosphere and all living creatures, is part of one great organism that is continuously evolving over a vast span of geological time. The central role of the health of the whole organism – the planet – is more important than the health of individual species.

The Gaia hypothesis predicts that the climatic and chemical makeup of the Earth are kept in balance for long periods by natural feedback processes which operate automatically. This system regulates our environment, keeping it suitable for life. It is almost totally self-sufficient, requiring only energy from the Sun. But if a significant change occurs, either internally or externally, the balance shifts as the whole super-organism tries to adjust.

The United States' space mission Apollo 17 took this picture of the Earth, showing the area from the Mediterranean to the great icecap of the South Pole. The Arabian peninsula appears at the very top, with the tip of Africa almost at the center of the image. Land and sea are vividly contrasted, and swirling patterns of clouds show different local and regional weather systems. One astronaut, seeing the Earth from space, commented that it gave him a stronger sense of the Earth's vulnerability than ever before – a sense that came, perhaps, from the unique experience of seeing the planet as a whole.

THE GLOBAL CLIMATE

THE Earth is the only planet that supports life. The first organisms appeared more than 3.5 billion years ago; today there are millions of species inhabiting the biosphere – the part of the planet where life is found. Climate is among the most influential factors that determine their survival.

The Earth relies on a constant supply of light and heat energy produced by the Sun, which is absorbed both by Earth's atmosphere and by the land and oceans below. The amount of heat that reaches the surface varies, creating climatic zones: tropical, temperate and polar.

The climatic zones are linked to areas of high and low atmospheric pressure. When air near the ground is heated, it expands and becomes less dense, creating an area of low pressure. The warm air rises, while cool air flows in to replace it. The result is a convection current of circulating air. The rising warm air cools as it gets higher in the atmosphere and increases in density. It begins to sink back down, forming a high-pressure area. Global temperature differences cause a circulation of air currents, with warm air rising from the tropics and moving toward the Poles, distributing heat energy along the way. In general, there are areas of low pressure over the Equator and the temperate regions; high-pressure areas cover the polar regions and the semitropical regions immediately north and south of the Equator.

The uneven distribution of rainfall is another important aspect of the different climatic zones, and it depends on both temperature and air movements. Tropical regions receive the greatest amount of incoming heat energy from the Sun, which evaporates huge amounts of water from the oceans and (to a lesser extent) the land. Warm air can hold more moisture than cool air, which tends to be dry – the warm air has a higher humidity.

As the tropical air rises, it cools and loses moisture, much of which falls back on the tropics as rain. The cooler, drier air then continues moving north, depositing little moisture in the warm temperate zone. As it passes over the warm land, it is warmed again, and more moisture is released in the cool temperate zone. By the time the air reaches the Poles, it is dry and cold.

Climate determines what vegetation can grow in a region, and how much. Tropical regions receive the most heat and moisture, and have the highest productivity – the total vegetation grown, usually measured in kilograms per square meter (kg/m^2). Rainforest produces about $3.5 \ kg/m^2$ per year, compared with less than $0.1 \ kg/m^2$ in deserts and polar regions. Productivity in deserts is limited by lack of water and in polar regions by lack of heat and light.

▷ African elephants walk at sunrise in a national park in Kenya. With the Equator passing through the center of the country, Kenya has a hot tropical climate. The soil was originally fertile, and large areas of grassland and woodland covered the region. However, much of the land in central Africa is now overworked and depleted. In this condition, stripped of vegetation, the soil reflects the strong solar radiation back into the atmosphere, preventing the formation of the clouds that bring rain. The entire region suffers from prolonged droughts.

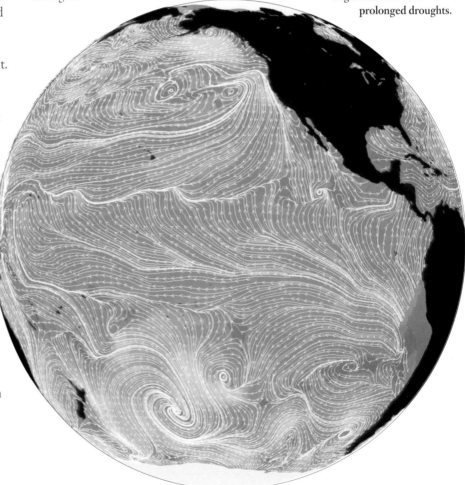

◁ The circulation of global winds has a major impact on ocean currents, which in turn affect climatic zones. Trade winds correspond to warm climates, westerly and easterly winds to cold and temperate climates.

Westerlies

Easterlies

Trade winds

△ A satellite map shows the direction and speed of winds over the Pacific Ocean. Trade winds, blowing north and south toward the tropics, are the dominant feature. Colors represent speed: blue is 0-14 kilometers per hour; pink and purple, 15-43 km/h; and red and yellow, 44-72 km/h. Winds in the South Pacific and Alaska are very fast due to storms in those areas, where cold and warm air meet.

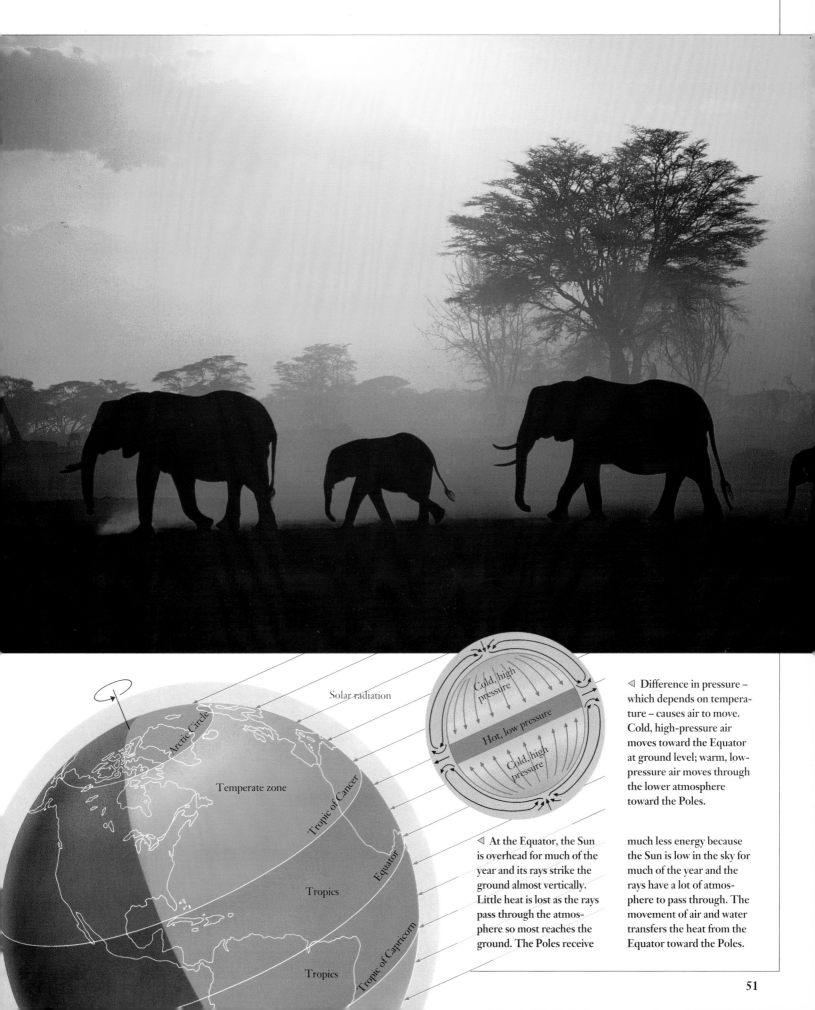

Solar radiation

Arctic Circle

Temperate zone

Tropic of Cancer

Equator

Tropics

Tropic of Capricorn

Tropics

Cold, high pressure

Hot, low pressure

Cold, high pressure

◁ **Difference in pressure –** which depends on temperature – causes air to move. Cold, high-pressure air moves toward the Equator at ground level; warm, low-pressure air moves through the lower atmosphere toward the Poles.

◁ **At the Equator, the Sun** is overhead for much of the year and its rays strike the ground almost vertically. Little heat is lost as the rays pass through the atmosphere so most reaches the ground. The Poles receive much less energy because the Sun is low in the sky for much of the year and the rays have a lot of atmosphere to pass through. The movement of air and water transfers the heat from the Equator toward the Poles.

51

WEATHER PATTERNS

THE daily pattern of conditions such as temperature and rainfall is called weather. It may change from day to day or even from hour to hour. Climate is based on the average weather conditions over a long period – about 30 years – and changes very slowly.

Weather begins with air movements. Warm air is less dense than cold air, so it tends to rise (convect), causing a reduction in atmospheric pressure. Wind blows from high-pressure to low-pressure areas. If the Earth did not rotate, cold, high-pressure air would simply flow from the Poles toward the Equator, while warm, low-pressure air from the Equator flowed toward the Poles. However, convection currents in the atmosphere are affected by the spin of the Earth on its axis. The warm, rising air spins more slowly than the Earth. Because of this, it moves along a curved path. In the northern hemisphere it is deflected to the right, and in the southern hemisphere it is deflected to the left. This is called the Coriolis effect. It also influences the direction of high-pressure systems (anticyclones) and low-pressure systems (cyclones). Anticyclones are associated with calm weather, cyclones with disturbances such as tropical storms.

At the Equator is a region of calm, warm, low-pressure air called the Doldrums. When it rises, it causes streams of high-pressure air to move toward the Equator from the north and south to replace it. These high-pressure streams are the trade winds. They do not blow directly north and south, for they are influenced by the Coriolis effect. The trade winds span nearly half the globe and dominate the weather systems of the tropics and semi-tropics.

Next to the zone of the trade winds, between about 30° and 60° latitude (the temperate zone), are the westerlies, the second major global wind system. Westerlies blow toward the Poles and are particularly strong in the southern hemisphere, where there is less large land mass to lessen their force. In the temperate zone, cold air from the Poles (called easterlies) meets the warm air from the tropics. These do not mix easily. The interaction of cold and warm currents and the spinning of the Earth result in high, fast-moving winds called jet streams. They are found 8-10 kilometers

above the surface of the Earth, travelling at 200 km/h. A jet stream is rather like a snaking tube that weaves an unsteady path, creating a series of areas of low and high pressure.

Unlike the Doldrums, which are a permanent fixture at the Equator, the large belts of winds – the trade winds and the jet stream – shift with the changing seasons as the Earth heats up and cools down. This is among the chief factors that produce weather conditions. For example, when the jet stream moves north, the air beneath becomes less dense and the tropical warm air can move north too, creating an area of low pressure. If it drifts to the south, denser, cold air moves south, creating a high-pressure area. The changing patterns of low- and high-pressure areas give countries in the temperate zones their changeable weather patterns. Hot, sunny weather results when the

△ A hurricane is a huge mass of swirling air, associated with banks of cloud, heavy rain and strong winds. Hurricanes form over tropical waters. The rising warm air draws in cooler air from the surrounding area, creating a huge storm system. This moves across the oceans, gathering energy. When it hits land, the winds can wreak havoc. All tropical storms are low-pressure cyclone systems.

jet stream moves north in summer, allowing warm tropical air to cover the land. When the jet stream moves south in winter, cold polar air moves south too, bringing cold winds and clear skies. Long-range weather forecasts attempt to predict the path of the jet streams because they have such a significant influence on the weather.

The trade winds move north and south over the Equator by about 5° – except over India, where they move by as much as 30°. This is partly due to the high temperature of the continent, which warms the air and creates low-pressure zones. These conditions are reinforced by shifts in the local jet stream, which moves to the south in the winter, bringing dry, high-pressure air down to India from the Tibetan plateau; in the summer it recedes back to the north, and the local low-pressure system resumes.

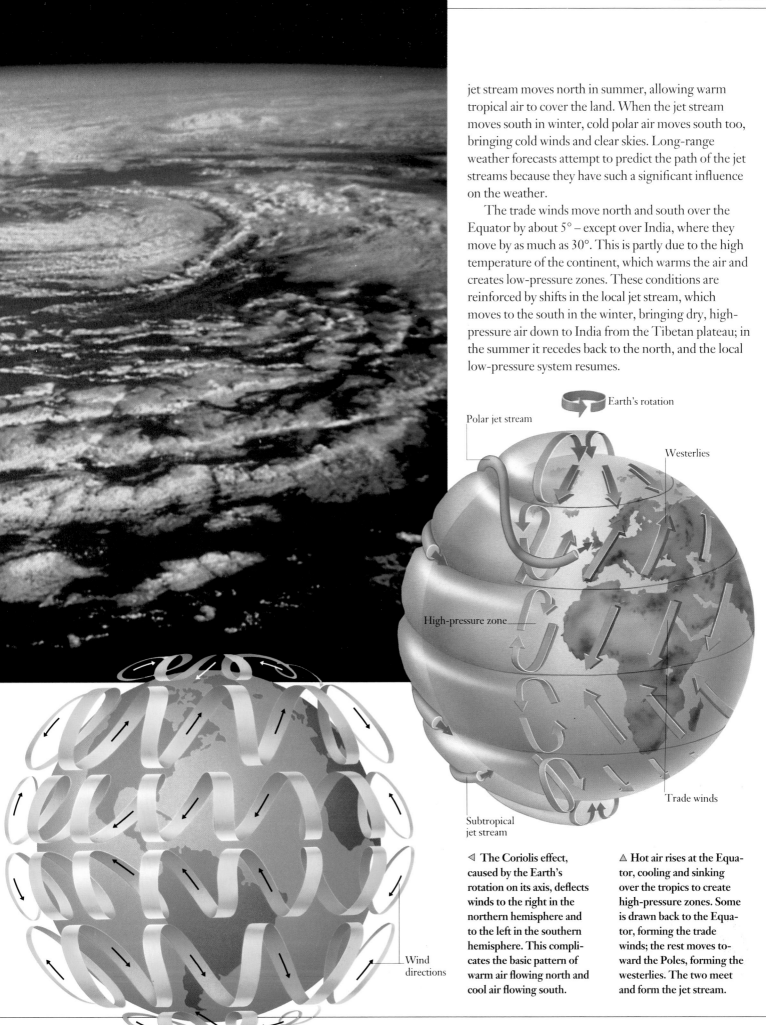

◁ **The Coriolis effect, caused by the Earth's rotation on its axis, deflects winds to the right in the northern hemisphere and to the left in the southern hemisphere. This complicates the basic pattern of warm air flowing north and cool air flowing south.**

△ Hot air rises at the Equator, cooling and sinking over the tropics to create high-pressure zones. Some is drawn back to the Equator, forming the trade winds; the rest moves toward the Poles, forming the westerlies. The two meet and form the jet stream.

MOVING OCEANS

MORE than 70 percent of the Earth's surface is covered by water, nearly all of which is in the oceans. The oceans were created millions of years ago when the Earth was relatively young. The cooling planet was surrounded by a layer of gases which included water vapor. Gravity prevented these gases from escaping into space. As the Earth cooled, its atmosphere could not hold all the water vapor; much of it condensed to form the five oceans: the Atlantic, Pacific, Indian, Arctic and Antarctic (also called the Southern Ocean). In parts of the Pacific Ocean, the water is more than 11 kilometers deep. At the polar regions the water is permanently frozen.

The oceans play a number of important roles in the biosphere. They act as a huge "sink" (absorber) for atmospheric carbon dioxide, much of which is incorporated in the skeletons of marine animals. They help to regulate the global climate by moving warm water from the Equator toward the Poles, and cold water from the Poles toward the Equator. The difference in temperature between land and ocean can produce climatic effects such as monsoons. Oceans are also the primary component of the water cycle, supplying huge amounts of water to the atmosphere to be distributed over the Earth as rain.

The water in the oceans is never still, because the oceans contain water currents much like the airflow in the atmosphere. Currents are set up by the effects of

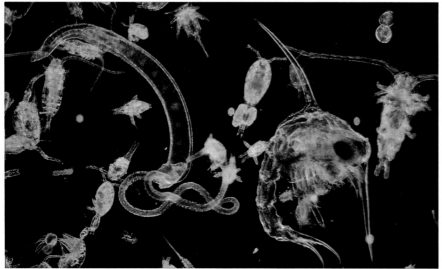

△ **Winds create surface currents, which are deflected by the Coriolis effect so that the water moves at a 45° angle to the wind. A sailboat rides the current, pushed by the wind.**

◁ **Plankton thrive in cold, nutrient-rich deep waters. These conditions are often created on the west coast of continents where currents and offshore winds push deep water up to form bands of cold, rich water in a larger area of warm sea. This is called upwelling.**

▷ **Major currents can change temperature and direction. The Humboldt current off the west coast of South America is normally cold and flows toward the Equator, but every few years it changes direction, travels south and becomes warmer. This is called El Niño, and it has a disastrous effect on the local fish population. It also affects the climate, disrupting rainfall and causing drought. The fires that raged in Australia in 1994 were a consequence.**

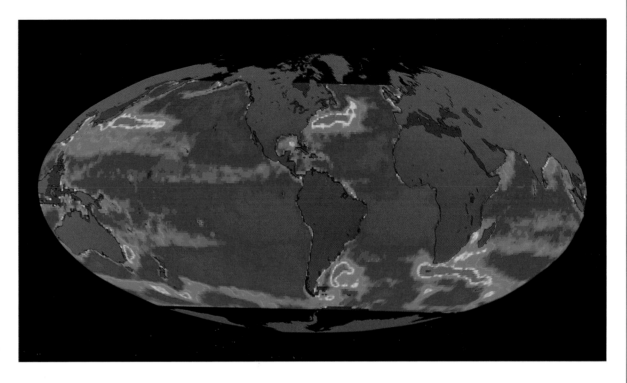

△ **The Gulf Stream (top center), the Brazil current (bottom center), the Kuroshio current off Japan (top left) and the Agulhas current off southern Africa (bottom right) are all warm currents, shown in red. Continents force the water to circulate within each** ocean. **Warm currents tend to form fast-flowing narrow bands and flow polewards on the western side of the ocean; cold currents are broad and slow. Deep-water currents move toward the Equator from the Poles, rising to the surface and joining the circulation.**

wind and by convection. The Sun over the Equator causes tropical waters to increase in temperature. Warm water is less dense than cold water, so it rises, causing convection currents. In the northern hemisphere, the warm currents move north to the temperate regions.

The opposite happens in the south. As the water cools, it sinks and moves back toward the Equator. These warm currents have a considerable effect on local climates. The Gulf Stream moves north from the Caribbean across the Atlantic to northern Europe, where it creates mild winters. Similar warm currents are important to areas of northern Japan and Alaska, where the climate would be much more severe in winter if not for a warm offshore current.

There are also cold currents that move toward the Equator from the polar regions. Cold currents, upwelling from deep seas, are rich in nutrients and support a rich and diverse community of plants and animals. An abundance of plankton attracts larger animals such as fish and whales. Recently, scientists have discovered currents within the deepest ocean trenches. The Humboldt current, for example, flows up the eastern coast of South and North America. However, every ten years or so, it turns warm, bringing heavy rain and rough seas, and wreaking havoc on wildlife and local fisheries. When this happens, it is called El Niño.

Surface winds cause waves to form. The size of the waves depends on the area of open water and the strength of the winds. Together, waves and tidal rhythms influence the shoreline communities.

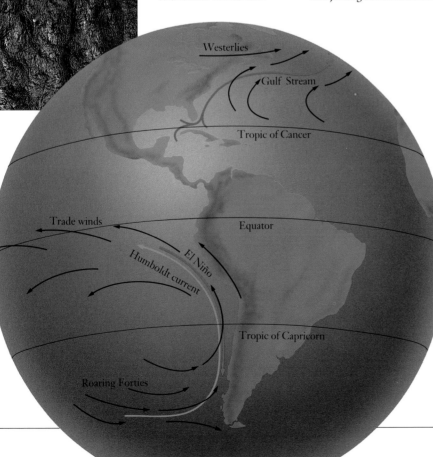

SEASONAL CHANGES

As the Earth orbits the Sun, it is tilted at an angle of 23.5° on an axis that joins the North and South Poles. Only one part of the Earth faces the Sun as it spins. This causes the phenomenon of night and day at different times across the globe.

Seasonal and daily changes are also governed by the way in which the Earth orbits the Sun. Most significantly, this affects the weather patterns of the temperate zones north and south of the tropics. For six months of the year, from May to September, the northern end of the Earth's axis is pointed toward the Sun. As the alignment of the axis approaches the Sun, the Sun appears higher in the sky in the nothern hemisphere, allowing more heat and light energy to reach the land, and increasing the temperature. On midsummer's day, the Sun reaches its highest angle, and the daylight lasts longest. For the other six months, from October to April, the southern hemisphere is tilted at the Sun. While the southern hemisphere experiences its summer, the northern hemisphere undergoes winter. As the Sun gets progressively lower in the sky, and less heat and light reach the ground, the days become shorter and colder.

In the tropics – which lie within 23.5° on either side of the Equator – the Sun is always relatively high in the sky. There is little seasonal variation in temperature and the air remains warm all year. Daylength remains almost constant throughout the year, with 12 hours of daylight and 12 hours of darkness. In the regions bordering the tropics, there are often distinct dry and wet seasons, though the temperature remains warm.

The Arctic and Antarctic are always very cold. In summer, the Sun never sets at the Poles, and there is continuous daylight. These few months are the warmest. In the middle of winter, the Sun never rises above the horizon and there is continuous night. One of the most dramatic seasonal changes occurs during the Antarctic winter. During the cold winter months, the size of the continent is doubled by the formation of sea ice, in some places extending more than 1000 kilometers from the land mass.

There are a number of seasonal winds. The mistral is a cold northerly wind that blows in the Mediterranean region during the winter months. Several hot dry winds blow from the Sahara throughout the same

▷ White storks pause on a rooftop in Spain en route from northern Europe to spend winter in Africa or the Middle East. Animals that migrate seasonally are programmed to recognize seasonal cues to leave and return. They navigate by a combination of magnetic sensing and ultrasound.

▽ Other animals, such as dormice, hibernate to avoid unfavorable seasons. They prepare for hibernation by eating extra food or by hoarding a supply.

▷ The tilt of the Earth as it orbits the Sun causes seasons in the higher latitudes. New seasons are marked by two equinoxes each year, when night 2 and day 4 are of equal length; and two solstices, one the longest day 1, the other the shortest day 3.

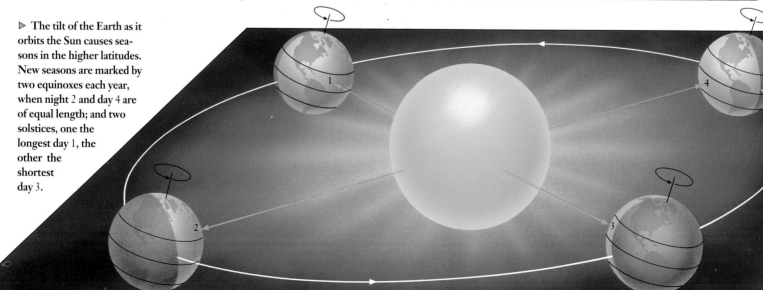

region between March and June. In North America, a warm dry wind blows down the eastern side of the Rocky Mountains in early spring. Called the chinook ("snow eater"), it causes a rapid melting of the snow. Many semi-tropical areas, particularly in southeast Asia, are affected by monsoon winds, which bring heavy rains in the summer but dry air in the winter.

Animals and plants have to adapt to these seasonal changes in order to survive. The change in day length, called the photo period, often provides the stimulus to change a pattern of behavior. Many birds migrate between regions of the world at different times of year in order to maintain a good food supply. Swallows, for example, live in Europe during the warm summer months. In autumn, as the days get shorter, they fly south to Africa, returning the following spring. Animals that cannot migrate often hibernate through the winter. They do this by reducing their heart rate and body temperature, and entering a resting phase during which they remain dormant through the winter months. The shortening days act as a trigger to prepare them for hibernation. They collect food and build up a store of body fat which must last them through the cold winter months. In the spring, many animals start their courtship rituals. Breeding in the spring ensures that their young are born in summer when there is plenty of food and they can fully develop before the arrival of winter.

Plants, too, are affected by the length of the day. Some plants flower only when the days are long, thereby ensuring that their flowers are pollinated by the available insects. Others flower when the nights are long. However, in the tropics, where seasonal effects are minimal, plants flower throughout the year.

▽ Seasons are reversed in the two hemispheres. In the northern hemisphere, spring starts at the March equinox and ends at the summer solstice, June 21. Summer lasts to the autumn equinox on September 23; autumn until the winter solstice on December 21. In the southern hemisphere, spring begins in September, summer in December, autumn in March and winter in June. Seasonal changes vary at different latitudes: the higher the latitude, the more extreme; the lower, the less so.

◁ The seasonal change of wind called the monsoon affects semi-tropical climates, mostly in southeast Asia. Hot summer weather creates low pressure over the land, attracting warm wet winds that bring heavy rain on which the growing season depends. In winter, cold conditions create an area of high pressure over the land and the winds blow out to sea. Farther south, in the heart of the tropics, rain falls more evenly throughout the year.

THE WORLD'S BIOMES

LIFE exists in only a small part of the Earth – in the lower atmosphere, on the surface and in the oceans. Together these form a single large ecosystem called the biosphere. Within it, organisms may live on land or in the water. These two different environments may be divided further into biomes, which are characterized by the kinds of plants that grow there.

Land covers less than a third of the surface of the planet, but 90 percent of all species live on it. The characteristic plants of a biome support typical groups of animals. This distribution of life forms around the globe is far from random. It is closely linked to the climatic zones – polar, temperate and tropical – because climate is a major factor that determines whether an animal or plant survives.

The two most important factors that influence which species can live in which land biome are the temperature and the amount of rainfall. These produce three main categories of terrestrial biome – grassland, forest and desert – in each of the climate zones. A forest, for example, may be cold (northern Canada), temperate (the Black Forest in central Europe) or tropical (the rainforest of central Africa). Biomes are often similar on different continents: African grassland looks like that of South America or Australia.

Biomes do not have precise boundaries, but blend into one another across broad geographical regions. Their patterns vary with climate, so that parts of the North African desert are at the same latitude as the

KEYWORDS

BIOME
BIOSPHERE
CLIMATE
POLAR
TEMPERATE
TROPICAL

▷ Salt-water biomes cover most of the planet, but most life is found on land. A map of the continents shows how terrestrial biomes are distributed in a pattern that corresponds to the climatic zones: polar, cool temperate, warm temperate and tropical. Varying amounts of rain contribute to the differences between the main types of terrestrial biome: forest, grassland and desert. The same type of biome may be found at different latitudes, and the same type of biome may be found in more than one climate. On a mountain, there may be a tropical biome at the base, forest on the slope and a polar biome at the peak.

▽ The Biosphere project covers 2.5 acres of land near Tucson, Arizona. A mini-rainforest BELOW LEFT and and ocean CENTER are two of the five complete biomes included in the custom-built glass and steel structure RIGHT.

THE BIOSPHERE PROJECT

In the Arizona desert, scientists have built a $30 million artificial biosphere to carry out experiments with self-sustaining ecosystems. Biosphere 2 contains the main natural biomes – rainforest, tropical grassland, desert, ocean – as well as two artificial systems, agriculture and urban.

The biomes contain 3800 species of plants and animals. The first team of eight humans, all scientists, spent two years inside the hermetically-sealed structure and succeeded in growing 85 percent of their food. However, there were problems. In the confined space, biomes are closer together than in nature, giving some animals access to an environment not their own.

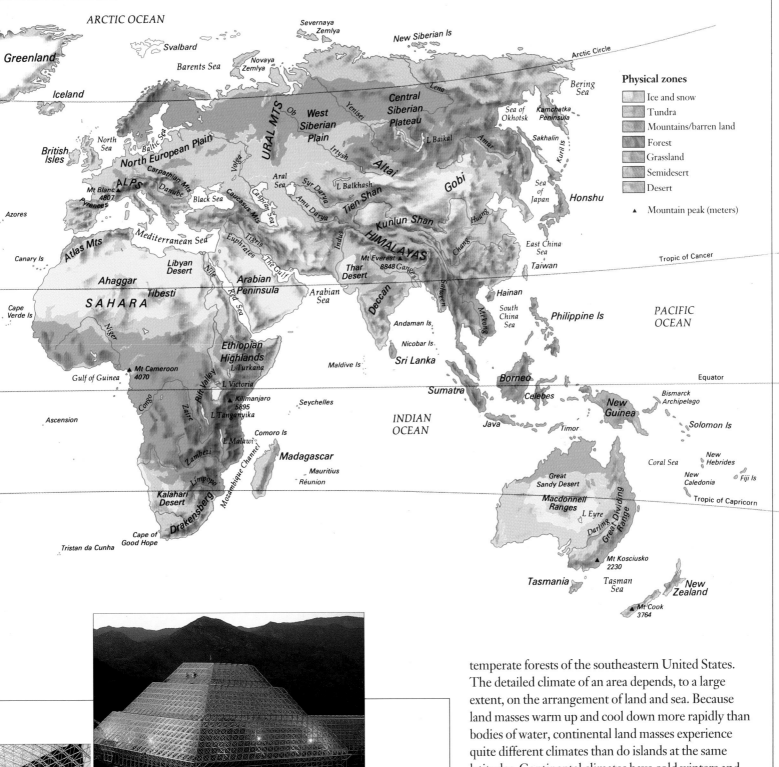

Physical zones

- Ice and snow
- Tundra
- Mountains/barren land
- Forest
- Grassland
- Semidesert
- Desert

▲ Mountain peak (meters)

The artificial climate is too moist; the desert biome has been overrun by grass and shrubs, while plants in the rainforest grew abnormally large. Plankton have not thrived in the ocean biome, and the coral reefs are starving. The humans began to have difficulty breathing as the overly rich soil consumed too much of the oxygen produced by the plants. More oxygen was absorbed by the concrete of the structure, which had to be painted to stop the absorption.

temperate forests of the southeastern United States. The detailed climate of an area depends, to a large extent, on the arrangement of land and sea. Because land masses warm up and cool down more rapidly than bodies of water, continental land masses experience quite different climates than do islands at the same latitudes. Continental climates have cold winters and warm summers, whereas maritime climates, which are influenced by the ocean currents, often have milder winters but wetter summers.

Climate is much less important in aquatic biomes. The single most important factor that determines what life can flourish in water is the amount of salt it contains. Salt-water aquatic biomes cover a much larger area of the Earth than any other biome, but they do not support as much life. Most marine life is found in the shallower water – down to about 200 meters in depth – over the continental shelves and slopes, where ultimately it depends on the abundance of plankton.

ON THE LAND

L AND-BASED (terrestrial) biomes supply most of the world's food and have a major impact on the global climate – forests, for instance, play a crucial role in stabilizing the oxygen/carbon dioxide balance of the atmosphere. Varying temperatures and amounts of rainfall produce the characteristic patterns of vegetation that characterize the terrestrial biomes.

Tropical rainforests grow in regions with strong sunlight, considerable rainfall and warm temperatures year-round. This biome is the richest on Earth in the diversity of its plant and animal life – it has as many as half the Earth's species. The soil is surprisingly low in nutrients, which break down quickly in the hot, moist environment and are rapidly absorbed by plants.

Temperate forests are found where there are distinct seasonal changes and moderate rainfall spread evenly over the seasons. Most of the trees are deciduous. Summer is the main growing season; in the winter, plants are dormant. Temperate forest is less productive and diverse than tropical forest, but it is home to animal species from birds to large predators. Because their timber is valuable, little remains of the temperate forests that once covered much of Europe and North America.

Boreal forest, or taiga, stretches across most of the world in a belt between the Arctic and temperate zones. The climate is cold, with long winters; water is plentiful, but frozen for much of the year. Because of the short growing season, boreal forest has little diversity of plant species, but these support a relatively large number of animals, from squirrels to grizzly bears. Insects and birds thrive in the short summers.

Tropical grasslands, called savannas, grow in areas with warm year-round temperatures and plentiful rainfall. In addition to grasses, there may be small trees and shrubs. Most of the world's hoofed animals live in savannas, along with a wide range of other animal species. Temperate grasslands cover large regions in the interiors of continents. They include the prairies of North America to the pampas of South America, the veldt of southern Africa and the steppes of central Europe. The rich soil and variable but mild climate make these regions ideal for agriculture, and in many places the original grass has been cleared for farming or by grazing of livestock.

Polar zone

High latitudes; ice, snow, tundra. Some shrubs and flowering plants, mosses and lichens. Boreal forest (taiga) at warm edge of zone.

Warm temperate zone

Lower mid-latitudes. Mixed grasslands; warm evergreen forests and aromatic shrubs; maquis and chaparral; desert.

Cool temperate zone
Mid-latitudes; coniferous (pine, spruce) and mixed broadleaf woodland; mixed grasslands.

Tropical zone
Low latitudes; rainforest and savannah; desert. High grasses (3 meters and above).

Polar biomes range from permanent ice and snow (in the extreme north) to tundra – frozen grassland that covers nearly 20 percent of the Earth's land area. Mosses, lichens and other low-growing plants form a thick layer over the ground, which thaws slightly during the scant two months of summer. Deciduous broadleaf forests such as this ancient beech wood in southern England are found in the cool temperate zone, along with other hardwood trees such as maple and oak. Large plains stretch across the temperate zone, from the steppes of Eurasia (cool temperate) to the pampas of South America (warm temperate); many have been cleared of their original grasses to make way for crops such as wheat and barley. A drier climate produces the semi-desert on the border of Arizona (USA) and Mexico, with cactus and cholla plants spreading their roots wide to trap water before it can evaporate. Both water and heat are abundant in the tropical biomes of savanna (grassland) and rainforest, where plant and animal life are at their most diverse.

Arctic grasslands, called tundra, are found at extreme northern latitudes but south of the polar ice at the Arctic Circle. The vegetation is dominated by low-growing perennial plants able to survive the cold conditions, harsh winds and dark winters. Temperatures range between –30°C in winter (which lasts six to ten months) and 10°C in summer. Most precipitation falls as snow. Topsoils are frozen for many months of the year, with only the surface thawing briefly in summer. The permanently frozen soil is called permafrost. The tundra is highly productive during the brief three to four months of summer and supports a rich variety of wildlife. However, most animals migrate or hibernate long before the worst of the winter weather arrives.

Deserts are areas that receive little rain, often less than 25 millimeters per year. Most are hot, and are found in tropical regions where air temperatures may rise to 50°C and surface temperatures to 90°C. Few plants can survive such arid conditions. There are also cold deserts such as the Gobi Desert in Mongolia and China. Parts of Antarctica, too, are technically deserts; one rocky valley has seen no moisture for 200 years, making it the driest place on earth.

AQUATIC BIOMES

THE Earth is accurately described as a watery planet because 70 percent of its surface is covered by water. Depth, temperature, the presence of salt and the movement of the water are the most important factors that distinguish the different bodies of water on Earth, and the life found in them.

Freshwater biomes are low in salt, and range from small seasonal ponds to permanent lakes and rivers. The mineral composition of the water depends on the surrounding rocks and soils through which the water has passed. Lakes and ponds are standing bodies of water that support life at four levels: the shore; the water surface; deep water ; and the bottom itself. Lakes that are low in mineral nutrients tend to be deep, clear and cold, with little vegetation on the shore. They are called oligotrophic lakes, and have small populations of plant and animal life. Warm, shallow, cloudy lakes with high nutrient levels are called eutrophic. They have diverse populations of fish and plants as well as plentiful shoreline vegetation. A third type, mesotrophic lakes, are between the two extremes.

Flowing water is very different from standing water. Because a lake or pond is not flowing, conditions are relatively uniform, whereas most rivers have different sections or stages. The upper course is usually fast-flowing and narrow, leading to a central section that is wide and slower in flow, and on to lower stretches with a much slower flow. The turbidity (cloudiness), strength of current, oxygen level, nutrient content and temperature all depend on the volume of water in the river and the profile of the land through which it flows. The flow of the water – like nutrient content, depth and temperature in lakes – affects animal and plant distribution. For instance, small animals such as plankton are not found in faster-flowing waters.

Fresh water flows from rivers down over the land to feed the oceans, which are the largest of the aquatic biomes – home to some 250,000 species. Oceanic biomes are classified according to whether they are coastal or open-sea, and by the depth of the water. Like lakes, they have upper and lower zones. The upper or pelagic depth zone can be further divided into three sections. The uppermost layer of water, the euphotic zone, extends to 100 or 200 meters below the surface, where plenty of light can penetrate, allowing

▽ **Water lilies float on the surface of a pond in a tropical climate. By not submerging its leaves, this plant makes effective use of the bright surface light rather than relying on the few wavelengths of light that can penetrate below the water.**

◁ Freshwater fish such as trout are adapted to swimming in fast-flowing streams and rivers, close to the water source, where the water is clear, cold and often turbulent. Trout require high levels of oxygen, which is taken in from the air and dissolved in the flowing water.

◁ A river has three distinct sections: a young stage with fast-flowing water; a mature river as it crosses level land, widening and laying down silt; and the late stage, in which the river flow is very slow and the amount of deposition has increased. The water in the first stage is the coldest and has the most dissolved oxygen. At the bottom of the course, the water is much more warm and still, almost resembling a lake.

△ Coral reefs are the marine equivalent of rainforests in their capacity to produce plant food and support other life. They are found in shallow waters with plenty of light and where currents bring a constant supply of nutrient and oxygen-rich water.

plants to grow and supporting a rich diversity of life, from whales to sharks and small fish. The bathyal zone extends down to 2000 meters, but no light can penetrate beyond 800 meters, and there is not enough to support plant growth. Beyond 2000 meters below the surface is the abyssal zone. Fish living in these waters have adapted to the dark and to the high pressure.

The lower major zone of the ocean is the benthic zone, which covers the ocean floor itself. Some marine animals have been found living close to hot water vents found in the deepest parts of the oceans. They depend on chemical energy rather than light energy from the Sun. A deep layer of nutrient-rich mud, made partly from the decomposing bodies of marine animals, feeds deep-living species such as sea slugs and brittle stars. Other nutrient-rich areas are found in regions such as continental shelves and where there are upwellings of water from deeper regions. However, most plant and animal life is found in coastal waters, which make up only 10 percent of the total area of marine biomes but contain 90 percent of marine life.

HABITATS AND NICHES

LARGE biomes, such as temperate forests, cover vast areas of the biosphere. Within a large biome are many smaller areas or microbiomes. These are usually referred to as the habitats of particular species of plants and animals. A habitat is a specific area with a characteristic set of conditions. It might be a pond, a woodland glade or a meadow. The groups of plants and animals that live in the habitat and interact with each other make up its community. The community depends on the nonliving components of the habitat, the abiotic factors, which include nutrients, temperature and water supply. These determine, to a certain extent, the type of animals and plants that can live in the habitat.

Even within a biome such as a rainforest, there are many individual habitats, each quite different from the next. There may be cool, low-growing, cloud-shrouded forest at high altitude and warm swamps by the rivers. In some places there may be openings in the canopy of the trees through which more light can reach the forest floor, creating a local microclimate with a different community.

Within a community are many different roles for the organisms to fulfill. Their diet, type of shelter, the range of temperatures they can tolerate, how much water and physical space they require, and whether they are active by day or by night are all aspects of this role, called a niche. Unless resources are abundant, a niche can only be occupied by one species; otherwise, species are in competition and some may not survive. All species therefore evolve and adapt to fill specific niches. Some niches are highly specialized and limit the type of habitat that the species can live in. In a rainforest, there is such diversity of species that many survive only by occupying thousands of specialist niches in different layers of the forest growth, from the forest floor to the highest branches at the top. These narrow niches allow many species in a small habitat.

The more specialized the niche, the more vulnerable the species is to any change in its habitat. A single species of bee may pollinate a single orchid species; their survival is linked. Species that occupy general niches, such as cockroaches, flies and mice, are more adaptable. Humans have occupied a general niche for most of our brief history, but have become more specialized in modern industrial societies that depend almost exclusively on a few limited resources.

KEYWORDS

ABIOTIC

BIOME

BIOTIC

COMMUNITY

COMPETITIVE EXCLUSION
  PRINCIPLE

HABITAT

MICROBIOME

MICROCLIMATE

NICHE

■ A woodland is a forest biome with widely spaced trees. It has an astonishing variety of habitats with varying amounts of space, light, heat, cover and moisture. The availability of open ground is important for large animals from deer through grazers such as bison (the European species are smaller than the North American), whose habitat encompasses the large area of the biome. Rabbits range over smaller territory and require relatively soft ground in which to dig their burrows. Birds build nests in the branches of trees and disperse seeds over the area. The same tree may be the site of nests for squirrels, which also live on the ground in some biomes. The damp, dark environment underneath fallen trunks provides habitats for fungi, worms, beetles, spiders and wood lice. Insects make nests in trees or on bushes. Frogs, fish and other species live in the stream that runs through the woodland; ducks and water rats at the edge of the water. These individual habitats increase the diversity of the woodland community.

WHERE LAND MEETS SEA

THE area where the land and sea come together is rich in life. Fresh water mixes with salt water in estuaries, where rivers empty into the ocean, creating aquatic habitats with lower levels of salt than the open sea. These diverse habitats – on mudflats, beaches and in salt marshes and swamps – are among the richest on Earth, equal to tropical rainforests in productivity. Their physical structures, such as dunes, also form barriers that protect the land along the coast from the ocean's extremes of waves, tides and salt water.

Rivers carry vast amounts of sediment and decaying organic matter which are deposited in the estuary, creating mudflats that stretch out into the sea. The water flowing over the mudflats changes as the tide rises and ebbs. It is a particularly demanding environment. Its inhabitants are subjected to periods of immersion, often with vigorous pounding by waves, followed by a period of exposure when they may be dried out by the air or attacked by land-based predators. The organisms living in such a habitat must be very adaptable. Although the conditions are difficult, there are many available nutrients, and mudflats contain millions of worms, shrimps and snails. Birds come at low tide to feed on these species.

As the mudflats increase in height, plants start to appear. In temperate zones, mudflats are first colonized by a small plant called a glasswort, and then by rice grass and other salt-tolerant plants. The level of the mud rises above the high-tide level. A salt marsh develops, crisscrossed by channels leading to the sea, and dominated by species of grass. Salt marshes typically attract huge numbers of birds to feed on the plants and on the small marine animals and insects.

In tropical regions, as mud accumulates on the seaward side of the swamp, it is colonized by mangroves, which advance toward the sea. The nutrient-rich mud supports an abundant and diverse community. Mangrove swamps are important coastal fish nurseries; the tangles of mangrove roots provide

▷ Waves and weather have cut ledges in cliffs on the northeast coast of England. They are ideal habitats for seabirds such as kittiwakes. The ledges are safe from non-flying predators, and the birds can land and take off easily. Kittiwakes' nests of seaweed are fixed to the ledges with mud and droppings. The young are genetically programmed to avoid careless falls.

▢ The dense growth of seaweed called kelp RIGHT in some coastal waters has been compared to a forest. Some species may grow up to 15 meters tall – growing a meter a day – in the nutrient-rich, bright water. Starfish BELOW are predators in tide pools. They have little tolerance of air, and are most often found in pools low on the beach where there is a plentiful water supply.

▽ Rivers deposit large amounts of silt into the ocean. Where they meet, the flow of the river slows and the silt is dropped onto the river bed. In time the silt builds up into mud banks. The mud flats expand and the delta extends further into the ocean, creating a typical triangular delta shape. Eventually the mud banks rise above the level of the tides and are colonized by plants. These plants, called halophytes, are very tolerant of salt water.

hiding places for newly hatched fish. When the tide goes out, crabs come to feed on the debris brought in by the tides.

Rocky shores provide a firm base for plants. They are colonized by marine algae or seaweeds and marked by distinct zones of life at different tide levels. Plants and animals that live below the low-tide mark cannot tolerate exposure to air. Kelps and oar weeds, which have fronds several meters in length, are found here. The middle shore is colonized by organisms that are adapted to exposure, such as the seaweeds called wracks, some of which can survive 80 percent dehydration. The upper shore is seldom covered by the tide. It has many land species adapted to a salt environment, with tolerant marine species such as sea kale and sea holly.

A sandy shore prevents seaweeds from attaching to it – the sand is always moving as the tide advances and recedes. Instead, communities of sand-dwelling animals such as crabs and worms develop. These animals hide in the sand during low tide, emerging to feed only when the shore is covered by water.

△ Salt marshes develop along the temperate coastlines. Their low mud banks are covered in vegetation and separated by channels through which the tide rises. Salt marshes attract flocks of birds which feed on the mud banks at low tide, seeking the worms and snails hidden the mud.

◁ Mangrove swamps develop along tropical coastlines. Mangrove trees are specialized plants that can survive immersion in salt water. Their roots form a raft to anchor the plant in the mud. Because the mud is low in oxygen, these roots grow up and out from it to absorb oxygen from the water at low tide.

2

CYCLES,
Chains & Webs

L IFE ON EARTH depends on two all-important natural
processes: the flow of energy from the Sun to the Earth's
surface, and the recycling of chemical elements such as
carbon and nitrogen found in all living things.

The Sun provides light and heat to the Earth. The atmosphere
causes about 40 percent of this to be reflected immediately back into
space, and another 15 percent to be converted to heat. To be useful
to plants and animals, the remaining energy that reaches the Earth's
surface must be transformed. The transformation begins with green
plants, which absorb solar energy and convert it to chemical energy
through photosynthesis. This is the beginning of the food chain,
on which the entire food supply on Earth ultimately depends.

Five chemical elements – carbon, nitrogen, oxygen, hydrogen
and phosphorus – and their compounds make up more than 95
percent of all living things. These elements are cycled continuously
between organisms and the environment. This process of recycling
also depends on the Sun: sometimes directly, as in the carbon cycle
(which involves photosynthesis), and sometimes indirectly, as in the
phosphorus cycle, which involves the breakdown of rocks by
weathering. The water cycle – the movement of water between the
Earth and the atmosphere – is driven by weather and climate.

Much of the Sun's warmth and light is soaked up as it reaches the Earth by the ocean, which nurtures a wealth of life forms in and above its upper waters. Once solar energy is converted into biochemical form and built into the bodies of living creatures, beginning with plants, it is passed from one species to the next through a complex of food webs. The ocean also provides many of the elements essential to life: its water evaporates to enter the atmosphere, to form clouds and rain, while carbon dioxide from the atmosphere dissolves in sea water, to be used by marine animals such as crustaceans to construct their shells.

ENERGY TRANSFORMATION

THE Earth depends on a continuous supply of heat and light from the Sun. A few organisms can use the heat energy directly – as when reptiles bask in sunlight to absorb heat – but the only organisms that are able to make use of light energy to make food are green plants and a few bacteria. The light energy captured by the green plants is converted to chemical energy, which can be utilized by animals. This can be thought of as a flow of energy from the Sun through an ecosystem.

When animals eat plants, they obtain both chemical energy and essential nutrients. Chemical energy is the "fuel" that drives the biological processes of life, whereas nutrients are the components that make up not only food but also living tissue itself. Both are passed from plant to animal or from animal to animal. Nutrients are chemical elements such as nitrogen and are part of the abiotic (nonliving) component of an ecosystem. Eventually, all nutrients are returned to the ecosystem and recycled. Energy conversion is less efficient. No animal can convert its total food intake into an equal amount of energy. Some energy is lost as heat and waste products and cannot be recycled back into the ecosystem. Decomposers, such as bacteria and worms, use the chemical energy and nutrients locked in the waste products and bodies of dead animals on which they feed. They play a key role in recycling nutrients through the system.

The conversion of energy in an ecosystem may be traced from one level to another. Primary producers (plants) are eaten by primary consumers (animals such as herbivores). These animals may in turn be eaten by other animals – secondary consumers – which, in turn, are eaten by tertiary consumers. The energy originally obtained by the primary producers is passed along, forming a food chain.

Each level in the chain is called a trophic level. The first level is occupied by the primary producers, and the second trophic level by the primary consumers. Consumers from different species may be at the same level in the chain, and share the same feeding habits. Plants are nearly always at the bottom of a food chain; therefore all animals are dependent, directly (in the case of herbivores) or indirectly (in carnivores), on plants for their energy requirements. An ecosystem can more easily support a large base of primary consumers than secondary or tertiary consumers, because the available energy decreases at each level.

KEYWORDS

AUTOTROPHIC
CONSUMER
DECOMPOSER
HETEROTROPHIC
PHOTOSYNTHESIS
PRODUCER

▷ Sunlight is the basis of the Earth's food supply, but its energy can be transferred to animals only through green plants (primary producers) such as grass. When animals such as rabbits (primary consumers) eat the grass, they gain chemical energy and essential nutrients. Rabbits attract predators such as owls, which are the secondary consumers in this ecosystem. Scavengers feed on dead and dying animals. All the wastes of the animals, together with the remains of the plants, are decomposed by microorganisms which recycle the nutrients back into the soil. At each stage there is a transfer of energy between plant and animal, or between animal and animal. However, up to 90 percent of this energy is wasted between one level and the next, usually in the form of heat loss. Eventually, all the energy entering the living components of an ecosystem will be lost in this way.

Solar energy

Plants

Heat loss

Energy flow

Herbivores

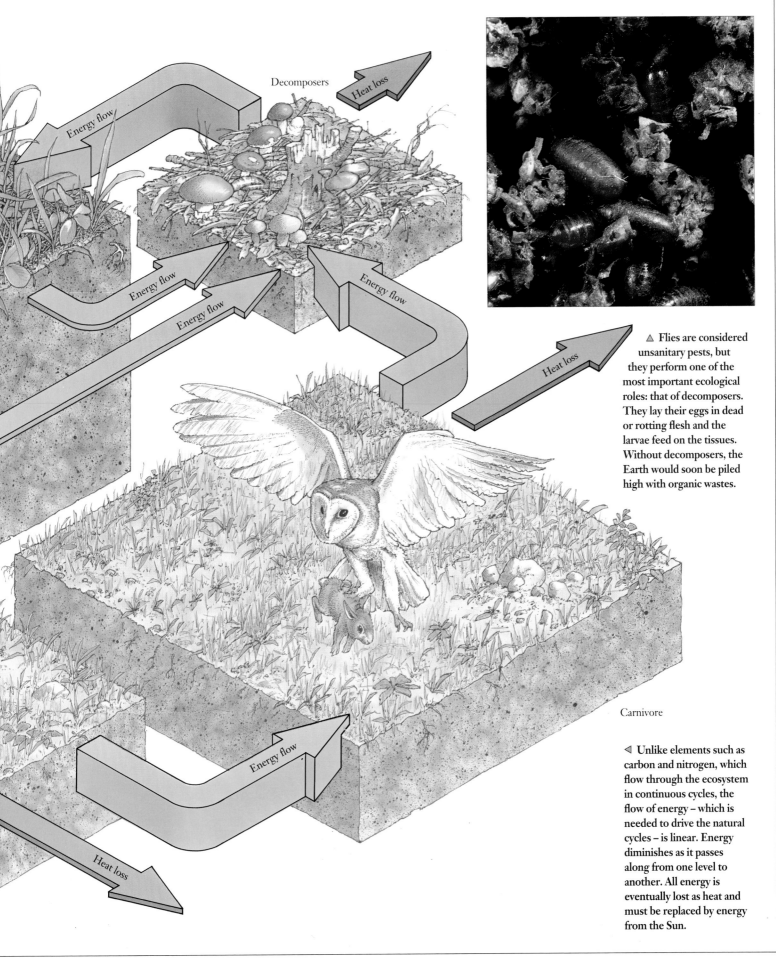

Decomposers

Heat loss

Energy flow

Energy flow

Energy flow

Energy flow

Heat loss

Carnivore

Energy flow

Heat loss

△ Flies are considered unsanitary pests, but they perform one of the most important ecological roles: that of decomposers. They lay their eggs in dead or rotting flesh and the larvae feed on the tissues. Without decomposers, the Earth would soon be piled high with organic wastes.

◁ Unlike elements such as carbon and nitrogen, which flow through the ecosystem in continuous cycles, the flow of energy – which is needed to drive the natural cycles – is linear. Energy diminishes as it passes along from one level to another. All energy is eventually lost as heat and must be replaced by energy from the Sun.

PRIMARY PRODUCERS

THE transformation of the Sun's energy into organic material is possible due to a green pigment called chlorophyll. Organisms that contain chlorophyll are commonly referred to as plants, although they also include cyanobacteria (blue-green algae) and algae such as seaweeds, along with the mosses, conifers and flowering plants that are the true plant species. All of these organisms are autotrophs, which means self-feeders. Alone among all forms of life on Earth, they are able to trap sunlight and use it to make food, in the process of photosynthesis. The rate of organic output from photosynthesis is called primary production, and plants – along with cyano-bacteria and algae – are primary producers.

In a marine ecosystem, the most abundant primary producers are single-celled algae, called phytoplankton, that float near the surface of the water. On land, it is the larger flowering plants such as trees and grasses that contribute the most to organic production. Terrestrial plants cover less than a quarter of the planet, but they are responsible for fixing 50 percent of the total sunlight captured on Earth by plants, and make up 97 percent of the Earth's biomass – the total mass of organic life.

Most plants appear green because their chlorophyll absorbs the red and blue wavelengths of light and reflects the green wavelengths. The absorbed light is used in photosynthesis, in which carbon dioxide and water are combined using light energy to produce oxygen and carbohydrates (sugar). The carbohydrates can be stored until they are needed. Up to half of the chemical energy stored by a plant is used for respiration – the biochemical process of breaking down organic compounds to release energy. The energy used in respiration cannot be recycled but disappears into the atmosphere. The remaining half is used for new growth and becomes available to animals that eat the plant. Photosynthesis makes another crucial contribution to the planet: one of its byproducts is oxygen, which animals breathe. The first photosynthesizing organisms, beginning some 2 billion years ago, allowed the Earth's atmosphere to become suitable for the life that now exists.

Plants are adapted to aid the capture of light energy for photosynthesis. Leaves are usually very thin, so that the carbon dioxide absorbed from the atmosphere does not have far to travel to reach the photosynthesizing cells. A large surface area helps to intercept a lot of light. Some leaves can alter their position to track

▷ The seeds of a plant contain an embryo and a food store protected by a hard outer layer. They remain dormant until conditions are favorable for growth. The life cycle of a bean starts with the release of seeds from the ripe pod of the parent. Water is absorbed through the coat, which expands and splits. The radicle (young root) emerges and takes root. At this stage the seed relies on a food store located in the cotyledons – embryonic leaves which are swollen at first, but shrink and wither as the plant grows.

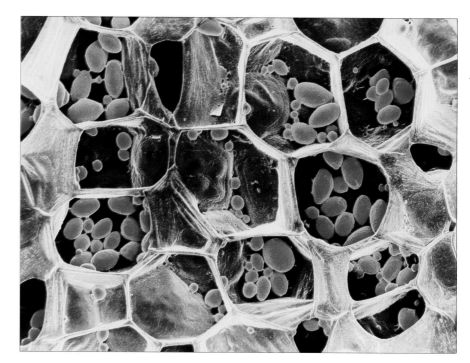

BROWN AND RED ALGAE

Only the blue wavelengths of visible light can penetrate deep water. To trap this light, some marine algae have brown and red pigments in addition to the green chlorophyll required for photosynthesis. The red and brown pigments mask the chlorophyll and give the algae their characteristic colors. Brown seaweeds such as kelps are found in the upper layers of water and can absorb green-blue wavelengths of light. Deepwater algae have red pigments that capture only the blue wavelengths of light found at greater depths.

△ Starch is made of many glucose units joined to form a chain, storing a large quantity in a small space. The chains form grains, found in specialized tissues in roots and seeds. Starch is stored in potato tubers and is used to fuel new growth in the spring. In seeds such as peas and beans it is laid down and used in the first weeks after germination. The size of the potato starch grains shown is 50 micrometers.

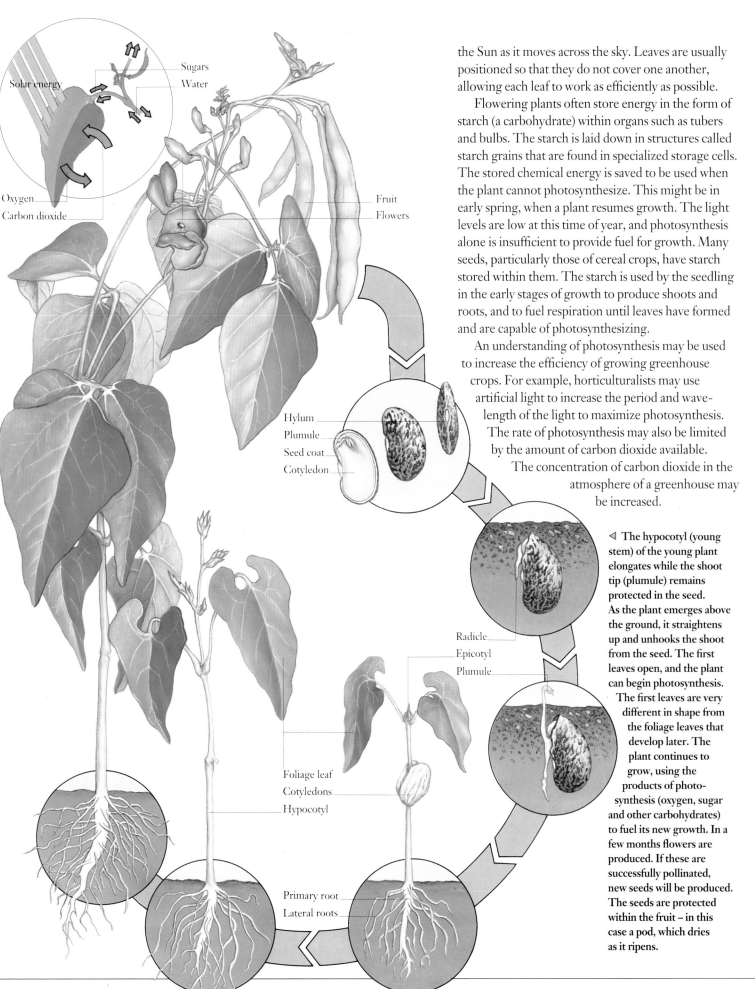

Solar energy

Sugars

Water

Oxygen

Carbon dioxide

Fruit

Flowers

Hylum

Plumule

Seed coat

Cotyledon

Radicle

Epicotyl

Plumule

Foliage leaf

Cotyledons

Hypocotyl

Primary root

Lateral roots

the Sun as it moves across the sky. Leaves are usually positioned so that they do not cover one another, allowing each leaf to work as efficiently as possible.

Flowering plants often store energy in the form of starch (a carbohydrate) within organs such as tubers and bulbs. The starch is laid down in structures called starch grains that are found in specialized storage cells. The stored chemical energy is saved to be used when the plant cannot photosynthesize. This might be in early spring, when a plant resumes growth. The light levels are low at this time of year, and photosynthesis alone is insufficient to provide fuel for growth. Many seeds, particularly those of cereal crops, have starch stored within them. The starch is used by the seedling in the early stages of growth to produce shoots and roots, and to fuel respiration until leaves have formed and are capable of photosynthesizing.

An understanding of photosynthesis may be used to increase the efficiency of growing greenhouse crops. For example, horticulturalists may use artificial light to increase the period and wave-length of the light to maximize photosynthesis. The rate of photosynthesis may also be limited by the amount of carbon dioxide available. The concentration of carbon dioxide in the atmosphere of a greenhouse may be increased.

◁ The hypocotyl (young stem) of the young plant elongates while the shoot tip (plumule) remains protected in the seed. As the plant emerges above the ground, it straightens up and unhooks the shoot from the seed. The first leaves open, and the plant can begin photosynthesis. The first leaves are very different in shape from the foliage leaves that develop later. The plant continues to grow, using the products of photo-synthesis (oxygen, sugar and other carbohydrates) to fuel its new growth. In a few months flowers are produced. If these are successfully pollinated, new seeds will be produced. The seeds are protected within the fruit – in this case a pod, which dries as it ripens.

THE CONSUMERS

ONSUMERS feed on plants or on other animals. Primary consumers, or herbivores, feed directly on plants. There are many different herbivores, including animals such as insects, reptiles, birds and mammals. They are adapted to feed on plant material, which can often be difficult to digest. Herbivorous mammals, for example, usually have flat grooved molar teeth to help them grind up their food. All plants have a rigid form due to cellulose in their cell walls. Herbivores must break down the cellulose to unlock the energy contained in the plant. Few herbivores can digest cellulose. Those that can include rabbits and ruminants such as cows and goats. They chew their food well before passing it on to the rumen (a compartment of the stomach), where it is fermented. Bacteria within the rumen secrete cellulase enzyme which breaks down cellulose. The digested food, together with some of the bacteria, then passes into a second stomach where digestion is completed. Rabbits have symbiotic bacteria in their large intestine that can digest cellulose. They also eat their food twice, once in the form of fecal pellets.

Herbivores are often the prey of secondary consumers, the carnivores (also called predators because many actively hunt their food) – a varied group that includes spiders, squid, fish, birds of prey, dogs and the big cats. In a food chain, there are always fewer secondary than primary consumers; fewer tertiary than secondary; and so on up through the top of the chain. The primary consumer is usually physically smaller than the secondary consumer, and obtains its food over a smaller area. The size of the predator and its range increases with the size of the prey.

Parasites are consumers found at all levels of the food chain. They may be primary consumers, such as parasitic fungi, and non-green plants that attack green plants. Potato blight is an example of a disease caused by a parasitic fungus. The leaves of an infected plant are the first to die, followed by the stem, with the collapse of the whole shoot. The fungus then feeds on the dead and decaying remains of the plant. Secondary and higher consumers are equally common among parasites, and include tapeworms, lice, ticks and many forms of bacteria, which may prey on animals ranging from small herbivores to humans at the top of the food chain. Parasites are the only consumers that live in or on the live bodies of the organisms that they feed on.

There is a surprisingly low level of energy transfer when one organism consumes another. For example, a herbivore rarely eats an entire plant and, because it cannot digest all of the cellulose, much of the food it eats passes through the gut undigested. Furthermore, some of the energy taken into the consumer's body is lost as heat energy during respiration. Ultimately, only a small amount of energy reaches the cells to be used to fuel new growth. When a carnivore eats a herbivore, there is an even greater loss of energy. Because of the inefficiency of the energy transfer process, and the huge energy losses that result, food chains rarely have more than four or five links.

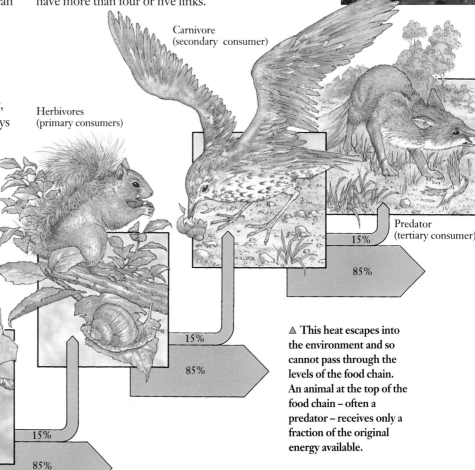

Carnivore (secondary consumer)

Herbivores (primary consumers)

Predator (tertiary consumer)

15%

85%

15%

85%

Solar energy

Plants (primary producers)

15%

85%

15%

85%

▷ The amount of energy that is transferred between organisms in a food chain is only 10–15 per cent. This small amount is due to energy wastage – up to 85 percent at each level, most of which is lost as heat when organisms respire.

△ This heat escapes into the environment and so cannot pass through the levels of the food chain. An animal at the top of the food chain – often a predator – receives only a fraction of the original energy available.

◁ Snakes are skilled predators and spend nearly three-quarters of their time hunting and digesting prey. They have acute senses to locate the prey; some have a heat-sensing mechanism that allows them to detect the presence of other animals, and special sensors to detect tastes or smells. The warmth or smell of the prey alerts a nearby snake, which strikes, either biting and injecting the prey with poison (if a viper), or squeezing the prey until it suffocates in the strong grip (constrictors or boids). When the prey is caught, the snake unhinges its jaw to enable it to swallow the prey whole. The largest snakes can swallow animals up to the size of small pigs and deer.

◁ Manatees are the only herbivorous mammals that live in fresh water. Also called sea cows, they "graze" mostly on grasses, occasionally on algae. Their forelimbs and snouts help them "pick" their food.

△ Monkeys eat mostly fruit, but some will eat almost anything available, from fruit and nuts to insects, snails, birds and fish. This Japanese macaque is eating flowers; it also eats seaweed.

PYRAMIDS AND WEBS

Most animals eat a variety of foods for a good reason: if anything reduced the supply of a particular food source, the consumer of that food would be affected by a shortage. By feeding on a range of foods, animals can avoid shortages. However, there are some animals that rely exclusively on a single source of food. For example, many of the sea birds of the North Atlantic around the Shetland Isles feed only on sand eels. Due to climatic change and over-fishing, the sand eel population has plummeted, and the sea bird populations have decreased dramatically as a result of starvation. This is an example of the interrelated feeding relationships that form a food chain.

There are different types of food chain. In a sequence in which plants are eaten by herbivorous animals (grazers), which in turn are eaten by carnivorous predators, the chain may be called a grazing food chain. There is also a detritus food chain, in which dead plants are consumed by decomposers. In a well-developed ecosystem, such as a forest, more than 90 percent of primary production is eventually consumed by the organisms in the detritus chain; less than 10 percent is consumed in the grazing chain. In contrast, in a less developed ecosystem such as a fishpond or farmland, 50 percent or more of production is consumed in grazing. Studies of grass-cow-human food chains in pasture farming have shown that future productivity depends on retaining at least 50 percent of the total annual production within the system; if less is retained, nutrient recycling and moisture cannot be maintained, and within ten years the ecosystem becomes severely depleted.

In the study of a particular ecosystem, it is possible to build up a complex food web that shows all the different feeding relationships. In such a food web, an animal may be both a primary and secondary consumer, feeding on both plants and animals; it may also be both a secondary and tertiary consumer, depending on which animal it feeds on at a given time.

A different way of illustrating a food chain is by using a pyramid of numbers. This display is designed to show the number of organisms at each feeding level. The pyramid begins with a large base and decreases sharply with each higher level because the number of

creatures is fewer in higher levels of the chain than in lower levels. In most cases, the pyramid is a reasonable representation, because there are always fewer carnivores than herbivores.

But the use of numbers alone to represent a food chain can be misleading. A large tree and a single small plant both count as one plant, even though there is an enormous difference in their sizes. For this reason, it is more appropriate to plot a pyramid of biomass – the mass of living matter at each level of a food chain. Sometimes the pyramid of biomass may be inverted: there is a greater mass of primary consumers than producers. This happens in marine food chains, in which zooplankton (consumers) often outnumber the short-lived, rapidly-reproducing phytoplankton (producers).

Pyramids of numbers and of biomass, however, can only provide a limited amount of information, because other critical information such as the availability of energy is missing. A pyramid of biomass does not indicate if any animals had large amounts of energy stored in their bodies as fat, or how much energy in the bodies of herbivores is passed on to carnivores. All this can be represented by a pyramid of energy. This shows the energy content of all organisms at each trophic level. It takes into account the number of organisms, their biomass and their energy content.

▷ Only a small amount of food is needed at the first level of a food web in an aquatic ecosystem such as a temperate pond, as long as the pond remains well supplied with light. A large base of primary consumers feeds on this small base, supporting in turn a surprisingly large number of higher consumers, up to the top of the food web. The first level of a land-based food web usually needs to be much larger.

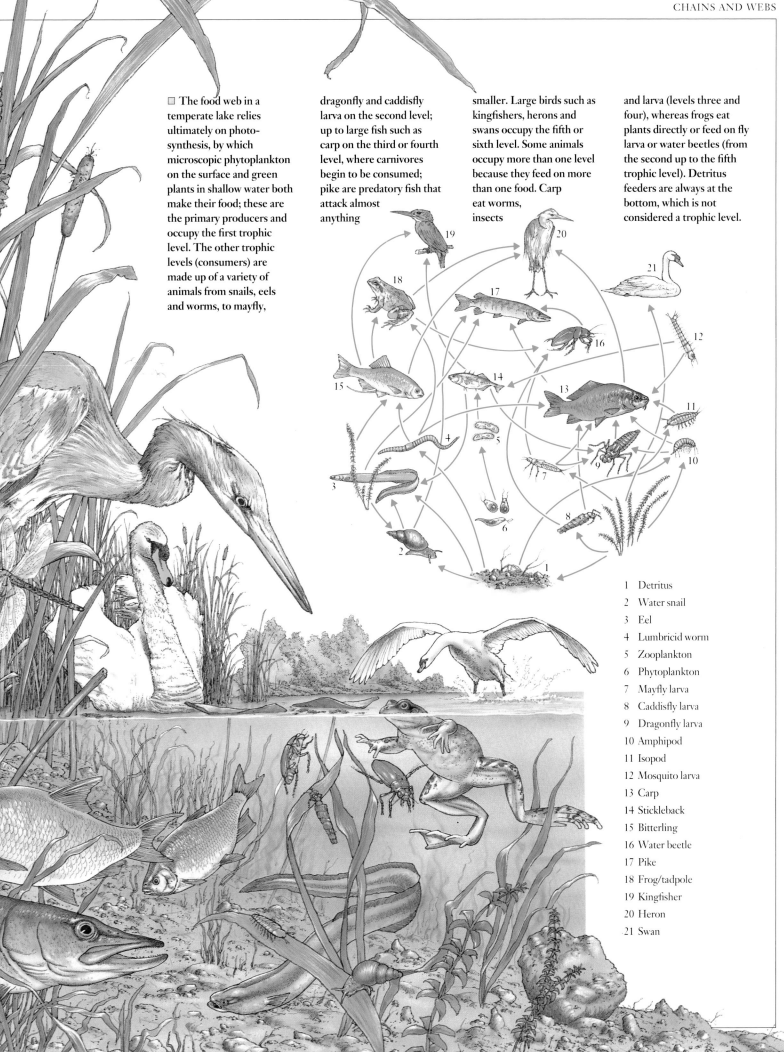

The food web in a temperate lake relies ultimately on photosynthesis, by which microscopic phytoplankton on the surface and green plants in shallow water both make their food; these are the primary producers and occupy the first trophic level. The other trophic levels (consumers) are made up of a variety of animals from snails, eels and worms, to mayfly, dragonfly and caddisfly larva on the second level; up to large fish such as carp on the third or fourth level, where carnivores begin to be consumed; pike are predatory fish that attack almost anything smaller. Large birds such as kingfishers, herons and swans occupy the fifth or sixth level. Some animals occupy more than one level because they feed on more than one food. Carp eat worms, insects and larva (levels three and four), whereas frogs eat plants directly or feed on fly larva or water beetles (from the second up to the fifth trophic level). Detritus feeders are always at the bottom, which is not considered a trophic level.

1 Detritus
2 Water snail
3 Eel
4 Lumbricid worm
5 Zooplankton
6 Phytoplankton
7 Mayfly larva
8 Caddisfly larva
9 Dragonfly larva
10 Amphipod
11 Isopod
12 Mosquito larva
13 Carp
14 Stickleback
15 Bitterling
16 Water beetle
17 Pike
18 Frog/tadpole
19 Kingfisher
20 Heron
21 Swan

3

CYCLES & *Energy*

T HE PROPORTIONS of the basic chemicals essential to life – carbon, nitrogen, water – in the biosphere have changed over geological time, yet they remain remarkably stable. Each one passes through the atmosphere, the Earth and through living organisms in a self-sustaining natural cycle.

Industrial and commercial agricultural activity can disrupt these cycles on a global scale. Even local disruptions can have disastrous effects on a regional level, polluting lakes and killing rivers, or causing deserts to form. Understanding these cycles and how we affect them is essential to the future of life on the planet.

One of the most serious consequences of disrupting a natural cycle is the threat of global warming. This may result from a greenhouse effect brought about by additional carbon dioxide in the atmosphere, derived primarily from the wasteful use of fossil fuels. Alternative approaches to energy are therefore bound up with the question of the sustainability of the major natural cycles. One approach is the exploitation of renewable energy sources, such as solar or wind power. Another, more controversial, is the development of a purely technological cycle, based on the element uranium, allowing us to unlock the energy bound up in its nucleus. This carries with it, however, a danger of devastating pollution.

Wind farms harness the Earth's circulating air to provide an inexhaustible supply of electricity. The search for renewable resources such as wind power is partly driven by the need to minimize human impact on the natural cycles. These "biogeochemical" cycles of crucial naturally-occurring chemicals play a key role in the maintenance of life on Earth. Atmospheric carbon, for example, is absorbed by plants and converted into all sorts of organic compounds. Eventually it is either released into the atmosphere once more as a byproduct of animal respiration, or locked up in rocks as fossilized hydrocarbons or carbonates. Humans have an impact on such cycles at many points, and may seriously disrupt them, with potentially dire consequences for the entire planet.

THE WATER CYCLE

ALTHOUGH Earth may be accurately described as a watery planet because so much of its surface is covered by oceans, water as a resource is far from abundant. This is because about 97 percent of the world's water is salt water, which is unsuitable for drinking, for irrigation, and even for many industrial purposes. Fresh water accounts for only about 3 percent of the total supply. Of this, less than 0.5 percent is readily available from rivers and lakes, which nevertheless provide 80 percent of the water used in industry and agriculture worldwide. The water supply may appear to be increased by natural means such as rain, or by artificial means such as drilling wells, but the amount available on the planet is constant. It is continuously recycled in the water cycle.

The oceans are the most important source of water, providing four-fifths of the total water in the cycle. Water evaporates from the surface of the oceans, leaving behind the salt. Some water also evaporates from rivers, lakes, the leaves of plants and the skin of animals as they sweat. The vapor rises in the atmosphere and cools. As it does so, it condenses to form water droplets high in the atmosphere. These

KEYWORDS

ATMOSPHERE
CLIMATE
CONDENSATION
DEW
GROUNDWATER
HYDROLOGY
IRRIGATION
LEACHING
OCEAN
TRANSPIRATION
WATER CYCLE
WATER TABLE

▷ Tropical rainforests absorb much of the abundant rain that falls on them, keeping the climate moist. Water is returned to the atmosphere when the trees transpire. The rest drains slowly through the soil into rivers, providing a year-long supply of water. Rainforests are also important in preventing erosion of the delicate soil.

▽ Snow is a crucial part of the water cycle, providing a third of the water used for irrigation worldwide. In the western United States, this figure rises to 75 percent: the flow of major rivers such as the Colorado and the Rio Grande is fed by melting snow in the spring.

droplets come together to form clouds. Their weight causes them to fall from the clouds as rain, snow or sleet – some of which falls on land, where it enters the next phase of the water cycle, and some over the oceans, returning the water to its source. On cold nights, the water vapor in the air near ground level condenses on cold surfaces such as glass of buildings and cars and forms on plants as dew. The fresh water in the rain runs off the land streams and rivers into the oceans, or evaporates into the atmosphere again.

A large part of the water that falls on the land as rain and snow penetrates deep into the ground, where it is stored in the spaces between rocks, called aquifers. This water is called groundwater. Aquifers feed underground springs and streams, which carry groundwater to the surface and to rivers, lakes and streams, where it evaporates or returns to the ocean.

Aquifers are replenished by precipitation falling on the land and sinking into the ground, but the process of water circulation underground is very slow compared with the process on the Earth's surface. Shallow groundwater may be recycled within a year, but in deep aquifers it may take thousands of years.

Disruption of the water cycle is becoming one of the major environmental problems facing humans. In heavily populated or heavily farmed areas, more water is drawn from both surface sources and aquifers than can be replenished. Depletion of groundwater in coastal areas leads to salt water seeping into fresh water underground and to subsidence of the land. Clearing large tracts of land increases surface runoff by removing vegetation that would otherwise hold some of the water in place. This in turn contributes to soil erosion and severe flooding.

▽ **Evaporated water –** mostly from the oceans but also from smaller bodies of water, soil, plants and animals – is deposited on Earth as rainfall, which may be absorbed by the ground or by plants, run off the surface, or evaporated back into the atmosphere. Runoff and evaporation take place quickly; water moving underground to rejoin the cycle moves very slowly. Without enough plant cover, more water runs off, causing damage in the form of soil erosion.

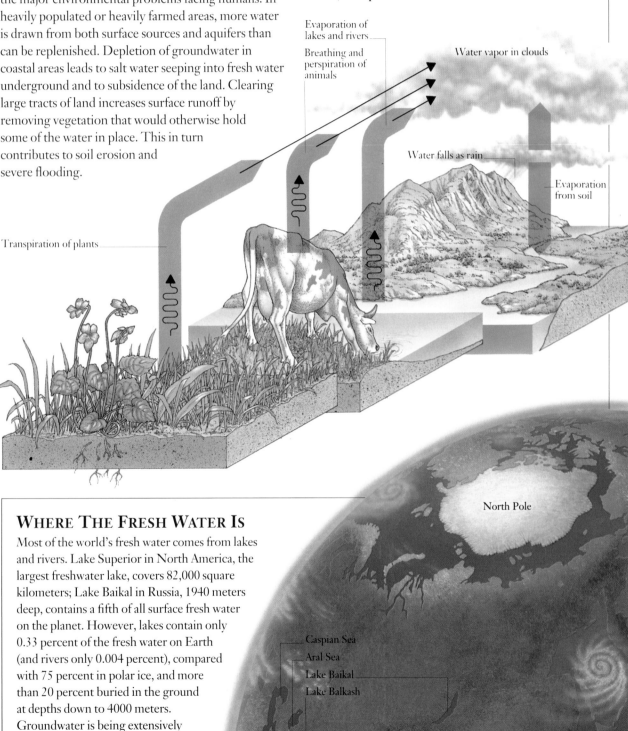

Evaporation of lakes and rivers

Breathing and perspiration of animals

Water vapor in clouds

Water falls as rain

Evaporation from soil

Transpiration of plants

North Pole

Caspian Sea
Aral Sea
Lake Baikal
Lake Balkash

◁ **The Aral Sea in Kazakhstan, central Asia, was once the world's fourth largest inland sea. Since 1960 it has shrunk 40 percent as the two rivers that fed it were tapped to irrigate cotton crops. The loss of the source water increased evaporation from the lake itself, which shrank some 27,000 km² in area. This huge seabed lies exposed, blowing sand and salt throughout the region. The local climate is now too dry to grow cotton.**

WHERE THE FRESH WATER IS

Most of the world's fresh water comes from lakes and rivers. Lake Superior in North America, the largest freshwater lake, covers 82,000 square kilometers; Lake Baikal in Russia, 1940 meters deep, contains a fifth of all surface fresh water on the planet. However, lakes contain only 0.33 percent of the fresh water on Earth (and rivers only 0.004 percent), compared with 75 percent in polar ice, and more than 20 percent buried in the ground at depths down to 4000 meters. Groundwater is being extensively tapped as modern demands outstrip the capacity of lakes and rivers.

THE CARBON CYCLE

CARBON is an essential element found in all living organisms. It occurs in organic substances such as carbohydrates, proteins, fats and nucleic acids such as DNA. Most carbon on Earth was originally released from the interior of the Earth in the form of carbon dioxide gas, which now makes up about 0.03-0.04 percent by volume of the air. This gas helps to insulate the Earth, retaining enough of the Sun's heat to keep the planet warm enough for life. Carbon dioxide is the most available form of carbon for living organisms. The process by which it is cycled around the ecosystem is called the carbon cycle. Molecules of carbon may pass through this cycle very quickly, or over millions of years.

Carbon dioxide is used in photosynthesis by plants to make carbohydrates, a vital source of food and energy. Carbon locked up in plant cells is transferred to animals that eat the plants. During respiration, all organisms break down food such as glucose into smaller molecules in the presence of oxygen, releasing energy at the same time. The product of respiration is carbon dioxide, released back into the air every time an animal exhales. Some of the carbon dioxide from respiration is used by plants in photosynthesis, while some of the oxygen made in

KEYWORDS

ATMOSPHERE
CARBON CYCLE
COMBUSTION
DECOMPOSER
FOOD CHAIN
FOSSIL FUEL
INORGANIC
ORGANIC
PHOTOSYNTHESIS
PLANKTON
RECYCLING
RESPIRATION

▷ **The food chain is an integral part of the carbon cycle. Green plants and algae, containing the carbon compounds they have made by photosynthesis, are eaten by animals, which incorporate the carbon compounds into their tissues as they grow. Both their living wastes and later their dead bodies are broken down by decomposers, releasing carbon dioxide. All living organisms respire, releasing more carbon dioxide back into the atmosphere.**

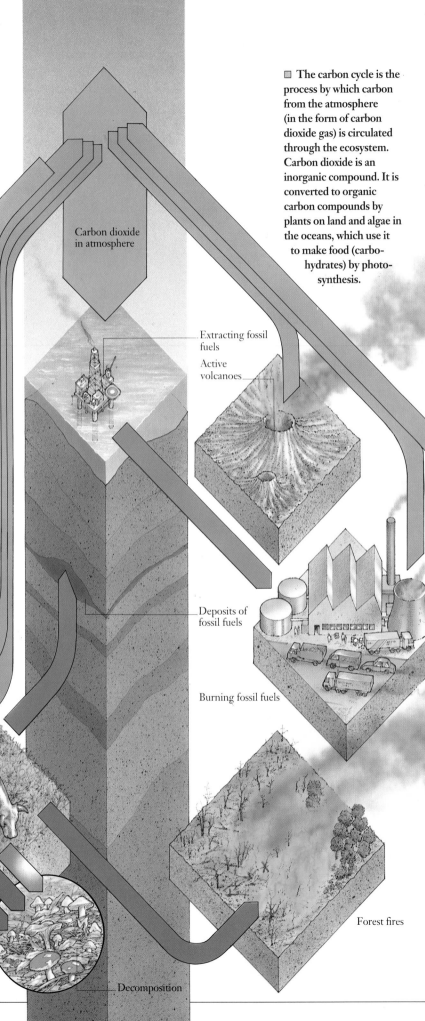

■ The carbon cycle is the process by which carbon from the atmosphere (in the form of carbon dioxide gas) is circulated through the ecosystem. Carbon dioxide is an inorganic compound. It is converted to organic carbon compounds by plants on land and algae in the oceans, which use it to make food (carbohydrates) by photosynthesis.

Carbon dioxide in atmosphere

Extracting fossil fuels

Active volcanoes

Deposits of fossil fuels

Burning fossil fuels

CO₂ for photosynthesis
CO₂ from plant respiration
CO₂ from animal respiration
CO₂ from soil organisms
Solar energy

Forest fires

Photosynthesis

Decomposition

▷ Natural fires in grasslands and coniferous forests release carbon dioxide into the atmosphere. This is due to the combustion of the carbon locked up in the cells of the plants. This carbon is oxidized in the presence of oxygen.

▽ Carbon dioxide is soluble in water, and huge amounts are dissolved in the oceans. The carbon is taken in by marine organisms and, together with calcium, forms calcium carbonate, which makes up most of the shells and heavy exoskeltons of marine animals such as mussels. When they die they sink to the seabed and become covered with sediments. Eventually, over millions of years, this becomes calcium-rich rock such as chalk or limestone.

◁ Natural processes such as volcanic eruptions and forests fires release carbon dioxide back into the atmosphere. Some of the carbon is locked up in the ground, eventually forming fossil fuels, such as oil, gas and coal. It returns to the atmosphere when the fossil fuels are burned.

photosynthesis can be used in respiration. For part of the day, the rate of photosynthesis exceeds the rate of respiration. This leads to an excess of oxygen, which diffuses out of the leaves. During the night, photosynthesis stops because there is no light. Respiration continues, however, so that a balance between carbon dioxide and oxygen production is achieved. It is for this reason that green plants are so important to the environment – particularly large groups of green plants, such as forests. They help to maintain the chemical composition of the atmosphere on which life depends.

When organisms die, their bodies are broken down by decomposers, which release more carbon dioxide. Sometimes the dead matter is covered by more dead and decaying material before it breaks down. In this way, over millions of years, fossil fuels – coal, gas and oil – have formed in the ground (including under the ocean). Carbon in this form remained locked up in the fuel deposits in the form of hydrocarbons and did not take part in the carbon cycle until the invention of modern devices such as the internal-combustion engine that made use of these fuels. When fossil fuels are burned, their combustion causes the carbon to oxidize (react with oxygen from the atmosphere), and carbon dioxide gas is given off. Forest fires also release considerable quantities of carbon dioxide into the atmosphere.

The Earth's oceans play an important role in the carbon cycle. Like forests, they are huge carbon "sinks", because their surface waters absorb vast amounts of carbon dioxide from the atmosphere. The carbon is used in photosynthesis by aquatic plants, or it may become incorporated into the shells of marine organisms as carbonate compounds. Decomposition and respiration also occur in the ocean, returning carbon dioxide to the atmosphere. Algae are as crucial to these processes as are green plants on land.

DISRUPTING THE CARBON CYCLE

ARBON dioxide is a greenhouse gas: it allows incoming short-wavelength radiation from the Sun to pass through the atmosphere, but blocks the reflection of heat energy from the Earth's surface out into space – rather like the glass of a greenhouse. Greenhouse gases form a natural blanket around the Earth, keeping it at a temperature that is warm enough to support life. There are several other greenhouse gases, but carbon dioxide is by far the most abundant.

Since the mid-19th century, the amount of carbon dioxide in the atmosphere has been slowly increasing, from approximately 280 ppm (parts per million) in 1800 to the present level of 350 ppm. Carbon dioxide levels are continuing to rise at an ever-increasing rate, and scientists have predicted that they could double over the next 100 years, disrupting the carbon cycle by exceeding the Earth's capacity to recycle this gas.

The increase in carbon dioxide has been caused by many factors. A certain amount comes from the burgeoning human population, which keeps an ever greater number of livestock – all of which produce carbon dioxide by respiration. Humans also burn wood and other fuels for domestic heating and cooking. But only since the Industrial Revolution in the 19th century has carbon dioxide production increased significantly. It is the fossil fuels used in power stations, factories and cars that contribute 80 percent of the 24 billion tonnes of carbon dioxide that are now produced annually.

△ **Destruction in the Brazilian rainforest, which lost 9 million hectares in 1987. Defoliation may account for 20 percent of the increase in carbon dioxide this century.**

▷ **Factories release millions of tonnes of carbon dioxide from the combustion of fossil fuels. Commercial use of other greenhouse gases, such as CFCs, is now subject to strict regulation.**

Some of the extra carbon dioxide is absorbed by green plants such as the trees of the tropical and temperate forests. This is why forests are sometimes referred to as the lungs of the world, because they help to balance the amount of carbon dioxide in the atmosphere. But the forests are being destroyed by commercial logging and clearing for agricultural use. As huge tracts are cleared, yet more carbon dioxide is released by burning or decaying trees.

Rising levels of carbon dioxide may affect the world's climate by causing weather patterns to change. The average global temperature may rise by as much

PLANTING TREES TO RESTORE THE CARBON CYCLE

Trees are often farmed as cash crops, but they offer far more value alive. Half a hectare of trees absorbs all the carbon dioxide produced by driving a car 40,000 kilometers. An environmental economist in the United States translated the 50-year lifespan of an average tree into $31,250 worth of oxygen and $62,500 in pollution control – compared with the $590 worth of timber it may be sold for.

Many countries have projects to plant more trees, such as the Green Belt Movement in Kenya (shown here) and the Chipko organization in India. It has also been proposed that industrial installations such as power stations should be built only if they are matched by planting sufficient trees to balance the amount of carbon dioxide they release.

Solar energy

Greenhouse gases
(CFCs and CO_2)

◁ **Greenhouse gases such
as carbon dioxide trap solar
energy in the atmosphere,
warming the planet.
Without them, the Earth
would be 30°C colder.
However, carbon dioxide
emitted by burning fossil
fuels is building up to
excessive levels.**

Atmospheric
pollution

Reflected heat

Melting ice
cap

Temperature (°C)

18

16

14

12

1800   1900   2000

Global warming

Local flooding
in estuarine
cities

△ **As increased
amounts of carbon
dioxide trap more heat in
the atmosphere, the
average global temperature
is rising: the Earth is now
0.3° to 0.6°C warmer than
it was 100 years ago, and in
the next 50 years it may
become 3° to 8°C warmer.
Cold climates may be most
affected, causing polar ice
to melt. This would raise
sea levels, flooding lowlying**

**land and possibly making
50 million people homeless.
With less ice, reflection at
the Poles would decrease;
more heat would be
absorbed, further raising
the temperature.**

as 3°C by the middle of the next century, making the
Earth hotter than at any time in the past 2 million
years. Climate changes have occurred before in the
Earth's history, but gradually. The speed of the
current global warming leaves no time for adjustment.
It has been estimated that plants need to move 90
kilometers towards the Poles for every 1°C rise in
temperature – a process that takes thousands of years.
Middle latitudes might experience longer growing
seasons and higher crop productivity, but also rapid
increases in the number of insect pests and decreases in
available water as the ecosystem went out of balance.
Major crop growing areas would shift to cooler areas,
where the soil might not be suitable for the new crops.
Extreme weather would become more common and
widespread; storms much more powerful. Tropical
diseases would spread to the former temperate zones.

THE NITROGEN CYCLE

NITROGEN is an essential element that all organisms need to function properly. Plants grown on nitrogen-deficient soils suffer stunted growth and early death. In animals, nitrogen is a component of crucial organic molecules such as DNA and proteins. Although 79 percent of the atmosphere is nitrogen gas, it is relatively inert and therefore cannot be used directly by most living organisms until it has been converted into nitrates or other nitrogen compounds. Certain bacteria in the soil, and cyanobacteria in the oceans, are among the few organisms that are able to carry out this conversion.

Nitrogen can be added to the soil as a result of electrical discharge during thunderstorms. The energy from lightning causes oxygen and nitrogen gases to combine with water vapor, forming weak nitric acid. This is washed down in rain and contributes to the nitrogen content of the soil.

Nitrogen is fixed by special nitrogen-fixing bacteria found in soil and water. These bacteria have the ability to take nitrogen gas from the air and convert it to nitrate. This is called nitrogen fixation. Some of these bacteria occur as free-living organisms in the soil. Others live in a symbiotic relationship with plants. Legumes such as clover, peas and beans have nitrogen-fixing bacteria in their roots which enable them to grow in nitrogen-deficient soil.

Nitrates taken in by plant roots are incorporated into large organic molecules, which are transferred to animals when they eat the plants. The wastes and remains of both plants and animals contain organic nitrogen compounds which are broken down by decomposers and converted into inorganic compounds such as ammonium ions. Nitrifying bacteria convert these compounds back into nitrates in the soil, which can be taken in again by plants and cycled through the ecosystem once more.

In denitrification, nitrates are converted back to nitrogen gas. Denitrifying bacteria are found in waterlogged soils where they release nitrogen gas, causing the soil to lose its nitrogen. Farmers normally try to prevent their fields from becoming waterlogged.

Because they are not readily available from the atmosphere, nitrates have been in short supply for most of the Earth's history. Nitrogen in artificially-produced compounds – the basic ingredient of fertilizers – is now more abundant than nitrogen from natural sources, and agricultural yields have improved

KEYWORDS

ALGAL BLOOM
BACTERIA
EUTROPHICATION
LEGUME
NITRATE
NITROGEN FIXATION

△ The roots of peas and beans have small swellings along their length. These contain millions of symbiotic bacteria, which have the ability to take nitrogen gas from the atmosphere and convert it to nitrates that can be used by the plant. These plants are often plowed into the soil after the growing season to improve the nitrogen content.

Atmospheric fixation

Atmospheric nitrogen

New nitrogen from volcanism

Fertilizer factory (industrial fixation)

Ammonium

Bacteria (denitrification)

Nitrite

Nitrous oxide

Nitrate

◁ **Modern crops need high levels of nitrogen, often applied as inorganic fertilizers. Traditionally, however, organic wastes are spread on the fields, as here. Consumer demand for organic produce, grown without artificial fertilizers or pesticides, has grown in recent years.**

▽ **Nitrates stimulate algae growing in water as well as plants growing in soil. If runoff from fertilizer gets into a body of water, algae grow so profusely that they form a blanket over the surface. This usually happens in summer, when the light levels and warm temperatures favor growth.**

dramatically. But the nitrogen cycle is easily unbalanced; even a small change can cause problems.

Modern crops, such as wheat and rice, require high levels of nitrogen to sustain their fast growth rates. The plants are harvested at the end of the growing season. The nitrogen within the crop is not returned to the soil, whose nitrogen level quickly becomes depleted. Farmers then have to add artificial sources of nitrogen – fertilizers – to the soil. The most common fertilizers are inorganic substances such as ammonium nitrate. Organic fertilizers, such as sewage sludge, manure and bone, are a good source of nitrate, but they can be expensive and are less convenient to apply.

Too much nitrogen can cause plants to become too lush and tall, so that they are more susceptible to damage from wind and disease. If a farmer applies too much nitrate fertilizer, particularly during wet weather, the water-soluble nitrate can leach out of the soil. It passes into water courses or soaks down to the water table – the supply below the Earth's surface. Eventually, the fertilizer ends up in a river or pond where it stimulates the growth of freshwater algae, which grow rapidly to form a green blanket over the surface of the water, called algal bloom. It can block the light to plants in the water, inhibiting their growth.

▽ **Atmospheric nitrogen is fixed industrially or naturally, by lightning or by soil bacteria that convert it to ammonium, then to nitrite, and finally to nitrates, which can be used** by plants. Nitrifying bacteria make nitrogen from animal wastes. Denitrifying bacteria convert nitrates back to nitrogen and release it as nitrogen gas. If too much nitrate gets into the water supply – as runoff from fertilizers, for example – it produces an overabundance of algae, called algal bloom. Algae have a short life cycle; when they die, they are consumed by increased numbers of bacteria, which respire and use up the oxygen in the water. This suffocates the more active aquatic animals such as fish. Water in this condition is called eutrophic.

Nutrient-rich water

Water treatment

Dead algae sink

Deoxygenated water rises

ENDLESS ENERGY

ENERGY derived from carbon-based fossil fuels – coal, oil and gas – accounts for 75 percent of the world's current energy use. At the current rate, oil and gas reserves will be exhausted in the next fifty years or so; coal will run out in several hundred years. Scientists have proposed alternative renewable sources of fuel in traditional power stations to reduce reliance on fossil fuels. In some schemes, wood is grown in specially planted coppices for burning in power stations. In Sweden and Norway, huge areas of land are now used for birch and willow, which are grown specifically for this purpose. They are referred to as biomass fuel.

Domestic waste is another potential source of energy. It is usually dumped in the ground in landfill sites. As a result of natural decomposition, methane gas is released. This gas has the potential to heat houses. Domestic waste can also be used directly by burning in industrial power plants, or it can be sorted first. Paper waste can be made into pellets and used in domestic boilers. This is called refuse-derived fuel. Burning waste for fuel reduces the need for landfill sites and generates electricity, but the fumes released must be cleaned to remove toxic gases.

Solar power is an effectively infinite resource. Solar panels can be used to heat a liquid, usually circulated

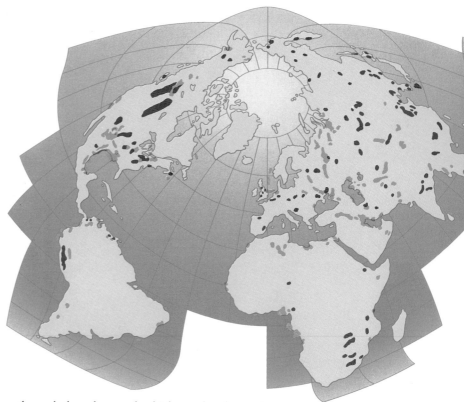

through the solar panel, which can then be used to heat offices and homes. There are even solar power stations where thousands of mirrors are used to direct sunlight into a central point where the intense energy of the Sun is used to heat up liquid sodium. This is used to produce steam, which spins a turbine, which in turn drives a generator to produce electricity. Solar cells convert light energy directly into an electric current. These are still relatively small and expensive to manufacture, though the technology is rapidly improving.

Wind farms have been a common sight in California and western England for many years. Wind turns thousands of propellers mounted atop metal masts; each directly drives a small electricity generator. Wave power is less well established, but may be useful in remote coastal locations where electricity is difficult to provide. In some coastal generators, the waves are funneled up a chute where they turn a generator. Open-sea designs use floats on the surface of the water. The floats rise and fall as the waves pass, and the reciprocating action rotates a turbine.

Falling water has been a source of power for hundreds of years. Large dams built across major rivers can provide hydroelectric power by passing falling water through a turbine to generate electricity.

Heat insulation

To soakaway

Cess pit for sewage

◁ **In India, as in much of the developing world, firewood is the major source of domestic fuel. Increasing populations mean greater demand for fuel, and for land for agriculture and grazing. This has led to scarcity in many regions, and people have to walk miles to find supplies. But with careful management, wood is a renewable resource.**

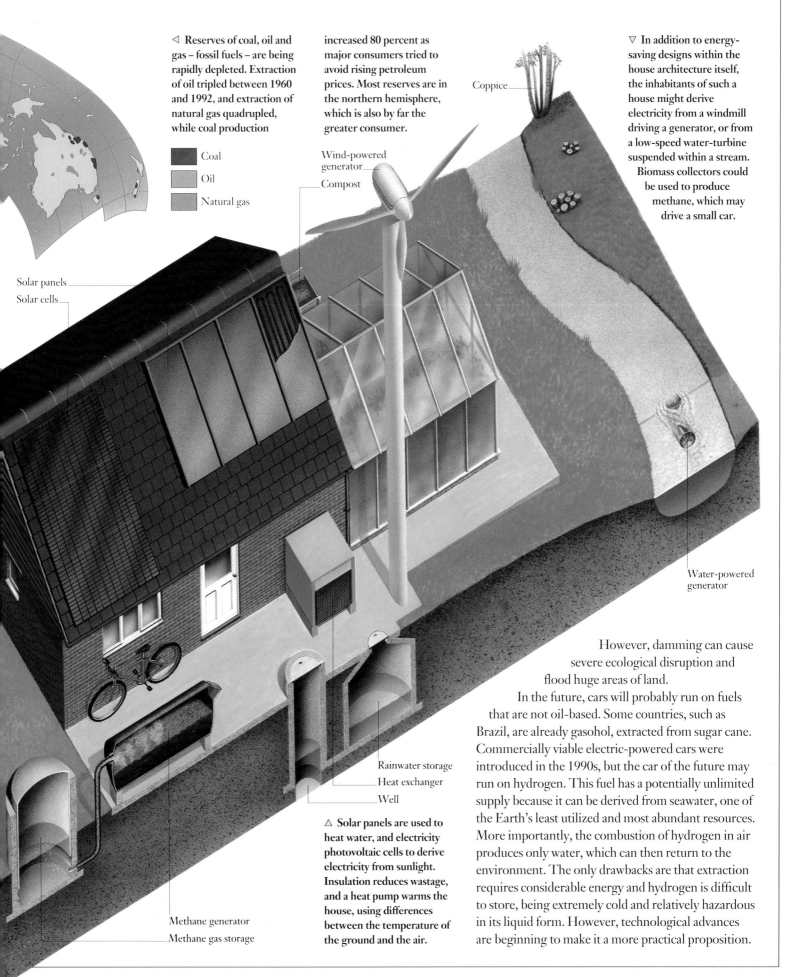

◁ Reserves of coal, oil and gas – fossil fuels – are being rapidly depleted. Extraction of oil tripled between 1960 and 1992, and extraction of natural gas quadrupled, while coal production increased 80 percent as major consumers tried to avoid rising petroleum prices. Most reserves are in the northern hemisphere, which is also by far the greater consumer.

■ Coal

▢ Oil

▨ Natural gas

Coppice

Wind-powered generator

Compost

▽ In addition to energy-saving designs within the house architecture itself, the inhabitants of such a house might derive electricity from a windmill driving a generator, or from a low-speed water-turbine suspended within a stream. Biomass collectors could be used to produce methane, which may drive a small car.

Solar panels
Solar cells

Rainwater storage
Heat exchanger
Well

Water-powered generator

Methane generator
Methane gas storage

△ Solar panels are used to heat water, and electricity photovoltaic cells to derive electricity from sunlight. Insulation reduces wastage, and a heat pump warms the house, using differences between the temperature of the ground and the air.

However, damming can cause severe ecological disruption and flood huge areas of land.

In the future, cars will probably run on fuels that are not oil-based. Some countries, such as Brazil, are already gasohol, extracted from sugar cane. Commercially viable electric-powered cars were introduced in the 1990s, but the car of the future may run on hydrogen. This fuel has a potentially unlimited supply because it can be derived from seawater, one of the Earth's least utilized and most abundant resources. More importantly, the combustion of hydrogen in air produces only water, which can then return to the environment. The only drawbacks are that extraction requires considerable energy and hydrogen is difficult to store, being extremely cold and relatively hazardous in its liquid form. However, technological advances are beginning to make it a more practical proposition.

THE URANIUM CYCLE

L IKE carbon, nitrogen and other naturally-occurring chemicals, uranium is an element found in the Earth's crust. However, unlike them it plays no part in biological processes, and in large quantities or concentrated form it is dangerous to living organisms. This is because uranium, which is the heaviest naturally-occurring element, undergoes radioactive decay, giving off radiation that can ionize the atoms of substances through which it passes. This can cause metabolic disorders in living cells, causing radiation sickness and, in many instances, cancerous growth.

The radioactivity of uranium was discovered in the late 19th century, and by the 1940s scientists had learned how to cause the uranium atom to split into approximately equal parts (fission) when bombarded by a neutron. This fission set up a chain reaction which released huge quantities of energy and if the uranium were packed sufficiently densely it would create a nuclear explosion. Alternatively, if the chain reaction is controlled, the heat can be released slowly and used to create steam for driving a turbine: this made possible the development of the modern nuclear power industry.

Nuclear energy is costly and involves complex technology, but it requires only small amounts of fuel:

▽ **The radioactive decay of uranium is a natural process that is an important source of heat inside the Earth. Mined uranium ore, though radioactive, needs enrichment and concentration before a fission chain reaction can be set up.**

Reprocessing spent fuel

Nuclear reactor

Fuel fabrication

Enrichment of fuel

Mining of uranium

△ **Rods of uranium are placed inside the nuclear reactor. The fission chain reaction is slowed down by lowering boron control rods into the reactor core and increased by raising them. The heat of the reaction is carried away from the core by a coolant (water, gas or pressurized water) and used to create steam outside the reactor.**

△ **The enriched uranium is packed into fuel rods, which are used in the reactor. Even an old fuel rod contains a valuable proportion of uranium, and it may be sent for reprocessing to extract this and recycle it into new fuel rods. Both the reactor and reprocessing plant create dangerous wastes.**

half a kilogram of uranium can give off as much heat as 1400 tonnes of coal. A proportion of that fuel can be recycled for reuse. The uranium cycle, although a technological cycle, is therefore an important component in considering the effects of industrial activity on the Earth's natural cycles.

Uranium ore is mined in Australia, France, North America and southern Africa. Less than one percent of the ore is uranium; the rest is left as spoil at the quarry. Uranium is found in two forms (isotopes): uranium-235 and uranium-238. Uranium-235 is more fissionable and is therefore better fuel; but natural uranium contains less than one percent uranium-235. The mined uranium therefore undergoes a process of enrichment to increase the proportion, by converting it into gaseous form, then separating the isotopes by centrifuge and diffusion, or by laser separation.

Enriched uranium is made into fuel in the form of rods which are placed in the core of a reactor, where they generate heat for about seven years, becoming increasingly less efficient as a porportion of the uranium decays into other elements, many of them (such as plutonium and strontium) highly radioactive and poisonous. The spent fuel rods may be taken to a reprocessing plant where they are dissolved in a strong acid and up to 96 percent of the remaining uranium is reclaimed for further use.

Nuclear power itself is a relatively clean source of energy; nuclear power stations do not emit air pollutants such as carbon dioxide, nitrogen oxide or sulfur dioxide. However, they produce a great deal of highly contaminating wastes. "Low-level" or slightly radioactive waste may be stored in drums and buried in shallow pits, but much of the waste from the fuel rods, as well as equipment in the reactor itself, may remain highly radioactive, dangerous and hot for hundreds or even thousands of years. This material must be sealed and stored so that there is no danger of radiation seeping into the ecosystem. It is often placed in stainless steel containers surrounded with a concrete jacket, or made into glass pellets and stored in steel drums. The nuclear industries are looking for geologically stable sites in which the drums of waste can be safely entombed for thousands of years. Equally, reprocessing is expensive and hazardous, and transport of spent fuel rods is also environmentally dangerous.

Burial for short-term storage

Long-term storage deep underground

▲ Spent fuel rods are highly radioactive. They are stored under water for two years to cool down and lose radioactivity. Water circulated around the rods absorbs the heat and acts as a barrier. Some fuel may be recovered by reprocessing; more often, the rods are stored at the plant. When space runs out, a new site must be found, and the hazardous cargo transported. This is one of the urgent problems associated with nuclear energy.

▲ Low and intermediate-level wastes are placed in steel drums and buried in shallow sites, although deep storage facilities are better to prevent leakage into the biosphere. The problems of long-term storage of high-level waste still have to be overcome. They require storage for 10,000 years, until their most radioactive contents decay. Solid waste is put in drums and stacked deep underground. These vaults are back-filled with concrete and sealed. It is difficult to locate suitable, geologically stable sites.

◀ The explosion at the Ukrainian reactor at Chernobyl in 1986 meant that the surrounding region had to be evacuated and sealed off for decades.

THE PHYSICAL *Environment*

SURVIVAL FOR ALL LIVING THINGS is a challenge shaped by their environment. The availability of food, living space, oxygen and water; the temperature; the soil type and many other factors are all physical factors of the environment. Plants and animals must flourish on what is around them, or adapt so that they are able to do so.

Physical factors such as soil, water and climate are known as abiotic factors. Soil contains nutrients as well as air and water. Its physical and chemical characteristics are influenced by the bedrock from which it is derived. Water, too, contains dissolved nutrients, but acidity, salinity and temperature are also important. Climatic factors include the amount of light, temperature and pattern of rainfall.

Biotic (living) factors are represented by the living organisms within a community. Their roles in the ecosystem may be as predators, parasites or competitors. Each has a different way of surviving, and each has a different effect on the system.

Abiotic and biotic elements are closely linked. On a global scale, they influence the biosphere – the part of the Earth that is suitable for life. On a smaller scale, they can influence the distribution of a particular species, affecting the makeup of a community and therefore the structure of an entire ecosystem.

From a human point of view, a wetland area – such as this group of tidal pools on the coast of Oregon in the United States – is often dismissed as wasteland: the ground is too wet for building, and the water is too salty for drinking. In fact, coastal wetlands are rich ecosystems with a great diversity of habitats. The range of species found in these areas depends on several factors. The most important is the amount of salt in the water. Others are the water temperature, the depth to which sunlight can reach (allowing photosynthesis to take place), and the availability of plant nutrients and dissolved oxygen.

WATER FACTORS

ALL forms of life need water in order to survive. The human body is about 70 percent water; other animals and plants range from 50 to 97 percent water. Living cells comprise a number of organelles and chemicals within a liquid, the cytoplasm, and the cell's survival may be threatened by changes to the proportion of water in the cytoplasm through evaporation (desiccation), oversupply, or the loss of either water or nutrients to the environment – a result, for example, of placing a cell designed for a freshwater environment into salt water.

Water is available very widely on the Earth, although in some desert areas the supply is limited, perhaps confined to a single rain shower annually. Water is naturally of variable quality, and the variations affect the type of organism that occupies a particular habitat. Apart from availability, the major natural variations in water quality are salinity, acidity, temperature, oxygen content and mineral content.

Because aquatic organisms are wholly surrounded by water, supply is not a problem. The suitability of the water depends on its temperature, oxygen content and salinity. Neither the salinity nor the temperature, of the water in the oceans vary greatly; for this reason it is not surprising that the earliest life was found in the oceans rather than on land. In oceans, and large lakes, the temperature of the water under the surface layer remains approximately 4°C in spite of the huge amount of solar energy absorbed during the summer.

In the oceans, however, differences do occur which affect the life within them. Currents, both cold and warm, move masses of water around the globe. Cold water holds more oxygen than warm water, and therefore supports a greater quantity of plankton and other life forms that feed off it.

In a smaller body of water such as a lake or inland sea, with little water movement, there are distinct layers of water which do not mix. The top layer tends to warm up quickly and can be much warmer than the layers below. However, in winter, this layer cools rapidly and may freeze, whereas the lower layers do not. The lower layers may have little oxygen.

On land, the availability of a regular supply of water is one of the most important factors affecting the presence – or absence – of plants and animals. The most species are found in the regions with the most

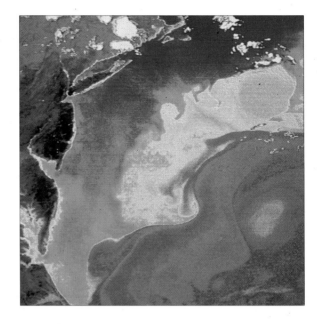

△ **Cold water has more oxygen and nutrients than warm water, but because it is deeper, it has less light and less life. At the shallow Grand Banks of Newfoundland, the warm Gulf Stream** meets the cold Labrador current which sweeps up the sides and creates a rich environment that supports billions of plankton. This forms the food supply for many fish and whales.

▷ **Most aquatic organisms can survive in salt water or in fresh, but not in both. Salmon are born in freshwater rivers, migrate to the ocean to mature, and return to fresh water to breed. They identify their birthplace by "tasting" the water. Moving from fresh water to salt water and back again requires a complete change in the physiology of their bodies, a feat that few other animals can achieve.**

abundant water. Few land organisms can drink salt water, as too much salt causes their cells to dry out. The temperature of the water is less important. It may be quite warm, as is the rain that falls in a tropical forest, or it may be cold or even frozen: some animals "drink" snow, and many rivers that are the main water supply for whole regions are fed by melted snow.

Most lakes contain fresh water, but in warm areas where there are high rates of evaporation, the water contains a higher-than-usual concentration of dissolved minerals washed from the surrounding rocks, forming a salt lake. These include the Great Salt Lake of the United States and the Dead Sea, between Israel and Jordan, as well as many of the lakes of the

▷ **Even a region with no rainfall can support a village if water from condensation is collected. In some parts of northern Africa tunnels are dug into hills with ventilation shafts to the surface. The tunnels are cooler than the air outside, so when air moves through the tunnels it cools and any water vapor in it condenses. A tunnel complex produces thousands of liters every day. This water is collected by the villagers.**

◁ Penguins are found in the cold southern oceans, particularly in the Antarctic. Although they live on land to breed and bring up their young, their terrestrial biome is utterly inhospitable and they have to take to the cold oceans to find enough food – mainly fish – to sustain them.

Shafts sunk to tunnel

Tunnels join up

Village water supply

Reservoir

△ The Namibian darkling bèetle collects water from drops of dew that form on their backs. They wait, with abdomens raised, on the top of desert sand dunes for fog to blow in from the sea.

Irrigation channels

Rift Valley of Africa. Their community of plants and animals is adapted to the abnormally high levels of salt.

The more regularly an animal needs to take in water, the closer it must live to the water source. Many birds can live far from water because their metabolisms are adapted to survive on only small amounts of liquid. In contrast, large animals tend to congregate around common watering holes. If rain is not abundant year-round, a regular seasonal supply makes an otherwise arid region more hospitable. The natural acidity or mineral content of the water reflects that of the surrounding soil.

Because they have so little water, deserts are occupied by fewer species than other biomes. However, occasional supplies of water from rainfall and flash floods may be sufficient for plants and animals adapted to a limited or irregular water supply.

CLIMATE FACTORS

LEVELS of light, temperature and the pattern of rainfall are the climate conditions that have the most significant effect on ecosystems. An adequate supply of each is necessary to support life. However, the range of 'adequate' is wide enough so that plants and animals are found in almost all climates.

Light from the Sun is critical for photosynthesis. Its intensity varies with the time of day or year as well as with the aspect of the habitat, such as whether it is on a north- or south-facing slope. The photoperiod is the number of hours of daylight in a 24-hour period, and it varies with the season. Many animals and plants in the temperate regions, where seasonal changes in day length are most marked, show strong responses to photoperiod. Plants detect the length of the dark period, and this affects the time at which many of them flower. There are plants that flower only when there are long days and short nights (the long-day plants); those that favor short days and long nights are called short-day plants. Day length affects the behavior of many animals and brings on dormancy, hibernation or migration. It also determines the breeding season. In an aquatic biome, the depth to which the light can penetrate controls the distribution of plankton and other life.

The functioning of cells is affected by temperature. Few organisms can grow if the temperature around them falls outside the range of 0–40°C. Freezing temperatures can damage living cells, whereas high temperatures alter the structure of body substances such as enzymes. Endothermic (warm-blooded) animals such as mammals and birds can regulate their body temperature in face of a changing external temperature. Ectothermic (cold-blooded) animals obtain heat energy from the environment. They spend much of the day basking in the Sun in order to absorb as much heat as possible, and hibernate during winter months to avoid the cold weather.

Temperature affects plants, too. In cold biomes such as the tundra and coniferous forests, plants are required to tolerate many months of sub-zero temperatures. Their growing season is restricted to the short, warm summer of 16 weeks or so. Snow covering these plants can act as an insulating blanket, keeping out the worst of the cold; the temperature below a layer of snow can be as much as 20°C higher than on exposed ground. At the opposite extreme, a few

▷ **Monarch butterflies undertake long migrations each year in response to seasonal changes, especially the changes in day length that occur in spring and autumn. In North America the butterflies move north as far as Canada during the summer. The offspring of these butterflies then migrate south again toward California and Mexico, where they spend the winter in communal roosts.**

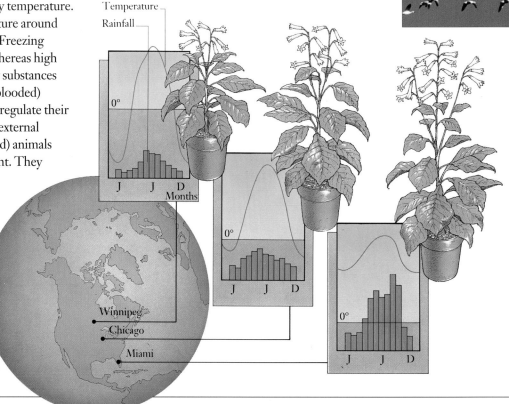

▷ **Tobacco can be grown from Miami (26° N) to Chicago to Winnipeg (50° N). The height of the plant depends on temperature, rainfall and nutrients, but its flowering occurs only when there is less than 12 hours of daylight. Tobacco was one of the first plants to be discovered to experience this phenomenon.**

■ Snow geese LEFT briefly make their summer home in the Arctic, but the extremely harsh winter conditions send them 3500 km (2000 miles) south to the Gulf of Mexico. BELOW Wildebeest on the savannas of Africa migrate with the seasonal rains in search of fresh grazing. The wet season from November to April is spent in the southern Serengeti plain; the dry season (May to July) in the western Serengeti; and August and September in the north. They move in huge herds together with zebras and some giraffes.

remarkable heat-loving bacteria, called thermophiles, can live in environments in which the temperature exceeds 100°C. Examples of these are found in areas in Iceland, New Zealand and the United States in which volcanic activity causes hot water to heat the surface where the bacteria grow.

Small differences in these conditions create micro-climates within habitats. For example, temperature and humidity vary significantly below and above a large stone, behind or in front of a tree, or one side of valley compared to the other. Air under the canopy of a tree is cooler and more humid than the air in an open glade. The north-facing side of a tree is more dark and damp than the south-facing side, and is often home to mosses and algae growing on the bark. The interior of some Arctic flowers can be 10°C warmer than outside due to their cup shape, which traps sunlight.

■ The dinosaur *Dimetrodon* BELOW RIGHT was an ectothermic (cold-blooded) animal, relying on the sun's energy to increase its body temperature. To make this process more efficient, the surface area of its body was increased by "sails" or plates along its back which lined up to intercept sunlight. The sails were made of elongated vertebrae linked by a sheet of skin, and could be turned away from the sun as *Dimetrodon* warmed up. Rattlesnakes ABOVE RIGHT are also ectothermic but do not have such elaborate anatomical equipment.

They bask in sunlight in warm weather, but as the days become colder and shorter they move under-ground to holes and caves, where they spend the winter in a dormant state until the temperature increases in the spring.

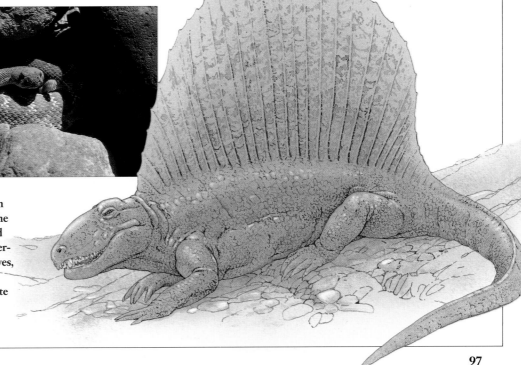

SOIL FACTORS

SOIL is a thin layer of material that lies over rocks, covering most of the land. It may be only a few centimeters deep, or it may extend several meters down to the rock below. Soil forms the link between the abiotic and biotic parts of a terrestrial ecosystem. Plant roots grow through it and take in water, minerals and oxygen. It has four main components: mineral particles (which may account for up to 60 percent), organic matter (about 10 percent), water (up to 35 percent) and air (up to 25 percent). The mineral particles of soil are derived from underlying rock as it undergoes weathering.

Physical weathering can be caused by changes in temperature that cause the rock to expand and contract, weakening it so that it eventually shatters. Plants such as mosses and lichens may grow through the cracks, loosening the rock material. Further breakdown occurs in chemical weathering as the rock is exposed to oxygen from the atmosphere or to acid in rainwater. Bacteria, fungi and lichens also contribute acids for chemical weathering.

The mineral particles in soil are distinguished by size: sand is the largest, then silt, and the smallest are clay. The proportions of each of these components give a soil its particular characteristics. A soil with a lot of sand and little clay is lightweight with many air spaces and drains easily, but it is poor in nutrients. A soil with mostly clay particles is much heavier, holds more water and is slower to drain. A loamy soil, with balanced amounts of sand, silt and clay, is best suited for agricultural use.

The character of a soil also depends on its chemical composition, which derives from the rock from which it formed. A sandy soil may also carry layers of iron or aluminum oxides (this is known as podsol); salty soils, with a high proportion of sodium and a clay-rich subsoil (solonetz), are frequently found in arid regions.

The organic matter that accumulates as the top layers comes from humus – dead material such as fallen leaves and the remains of animals. Humus gives soil its dark color and nutrients, and improves water retention (in sandy soils) and drainage (in clay soils). Bacteria in the humus play an important role in fixing atmospheric nitrogen and making it available to plants. New soil contains no humus. Mature soil takes as long as 10,000 years to develop, while plant cover grows to allow nutrients to circulate between the soil and vegetation. If it is not overexploited, the soil remains fertile for millions of years. Without plant cover, the soil becomes badly eroded within decades and cannot be replenished.

▷ **Soils develop layers called horizons. Most mature soils have at least three; there may be as many as six. The topsoil, the top layer, is rich in organic matter. Water carries minerals down to the lower horizons in the process called leaching, which occurs mainly between the topsoil and iluvial layer. The subsoil, below the iluvial layer, is paler and has less organic matter. It is derived from the bedrock below. Soil particles are separated by air spaces that contain varying amounts of water, oxygen and nitrogen. Oxygen is found mostly in the upper layers (topsoil) and is used by roots for cell respiration. Nitrifying bacteria make nitrogen available to plant roots. Other organisms in the soil community include mites, spiders, slugs, fungi, centipedes and worms. Soil types are classified on the basis of their texture, chemical composition and organic content.**

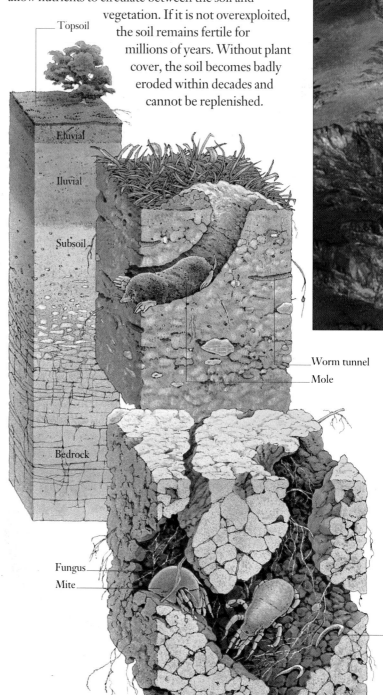

Topsoil

Eluvial

Iluvial

Subsoil

Bedrock

Worm tunnel
Mole

Fungus
Mite

Nematode

◁ Although rainforests are abundant with plant life, their soils are surprisingly delicate and low in nutrients. This is because the plants take in nutrients so quickly that the majority of nutrients are locked up in the living biomass rather than in the soil. When tropical forests are cleared, the soils are exposed to heavy rain. Without vegetation to retain it, the water runs off the land, carrying soil particles and leading to extensive erosion. The dangers are starkly evident here in Madagascar. Restoring such land once it is damaged is all but impossible. Tropical and temperate soil is renewed at an average rate of 3 cm every 500 years (though it may be twice as slow); erosion occurs 18 to 100 times faster.

ACID AND ALKALINE SOILS

The varying levels of minerals and acidity in the soil have a considerable effect on the types of plants that are able to grow. An acid sandy soil, low in nutrients, is favored by coniferous trees and by plants such as heathers RIGHT, which cannot tolerate much calcium; these are called calcifuges. If calcifuges are grown in alkaline soils, they suffer from poor iron metabolism. In contrast, calcicoles (calcium-seekers) grow in calcium-rich alkaline soils often found in grasslands. Typical calcicoles are the grassland plants growing on thin chalky downs FAR RIGHT.

Calcium and other alkaline compounds may accumulate in the soil in arid climates. They may be leached (washed) away by irrigation, as long as drainage is adequate to prevent waterlogging.

SUCCESSION

A HABITAT such as a pond or an area of grassland does not stay the same for ever. If left unmanaged by humans, it will change, and new species may move in, replacing the existing species. Each new species that arrives contributes to further changes in the habitat, so that it becomes increasingly different from its original state, as well as more complex.

Ecological succession describes the process by which this occurs, and the whole succession is called a sere. Each of the stages is characterized by plants and animals that are adapted to the prevailing conditions. For example, a lithosere is a succession of plants that colonize bare rock. This stage is called primary succession. It may occur after the land has been stripped of soil by a disaster such as a volcanic eruption, earthquake or flood, or by human activities such as surface mining. Algae and lichens are among the first species in the lithosere, which are called pioneer species. They contribute to the breakdown of the rock and the formation of soil. Larger plants such as mosses and ferns can then grow, further altering the soil and other conditions so that still larger and more complex plants can become established. In this way, the habitat develops into a mature, stable and self-sustaining community, called a climax community.

This progression can be seen on temperate grasslands, which are normally maintained by grazing animals, such as rabbits, that remove the shoots of trees and shrubs. Many of the open grasslands of the

▷ Not all succession takes place over a period of time. Some is evident over distance – for example, on a mountain. As altitude increases, both the climate and the vegetation change. The higher up the mountain and the colder the climate, the fewer species there are, the smaller plants become, and the more slowly they grow. Forest may grow on the lower to middle slopes, but the top of the mountain (called the alpine zone) is treeless. Some mountains may have tropical forest at their base and snow at their peak. The pattern produced by this succession is called zonation.

▽ Millions of trees fell in a hurricane that struck England in 1987, providing an opportunity for secondary succession. In a few seasons, the area will be covered by plants such as bluebells and foxgloves. The extra light and lack of competition allows tree seeds buried in the ground to begin to grow. Trees will again take over the woodland, which will reach maturity in about 300 years.

world are managed to keep them at this stage. If the animals are reduced in number or removed, larger plants (shrubs) replace the grass. The larger plants attract different groups of animals. Later, tree species such as oak and beech arrive. They dominate the vegetation and create a climax woodland that is stable for hundreds of years – until it is cut down, or felled by a storm, a fire or disease. Then, as long as the soil is intact, succession begins again. This type of succession, called secondary succession, is common on abandoned farms and in forests that have been cut down. However, some species in succession compete for resources rather than facilitating the arrival of new species. Crab grass, found on abandoned farms in the

▽ A pond is not a permanent open body of water, but changes over a period of time. 1 As organic matter builds up on the pond floor, marginal plants such as reeds may start to colonize the water and extend towards the middle. Their roots cause more buildup of mud, and the depth of water gradually decreases. 2 Plants that need deep water disappear and are replaced by those that prefer shallow water and can survive periods of drought in summer. They spread into the middle of the pond and it becomes a temporary wet area in rainy weather. As the pond dries out, the shallow-water plants are replaced by plants that prefer damp soil. 3 Eventually the soil is dry enough for trees and shrubs; the pond has now become a woodland.

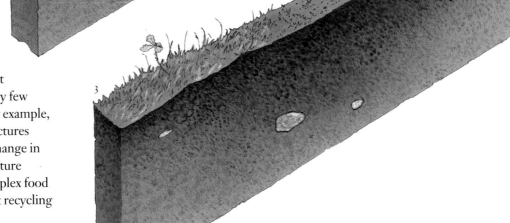

southeastern United States, is an example. The seeds of pine trees require bare land to grow; to do so, pine trees accumulate litter that kills off many other plants.

In the early stages of succession, an ecosystem has few species. Its food chains consist mostly of producers and herbivores, with very few decomposers. It is not yet self-sustaining; for example, it has not developed the roots and other structures necessary for nutrient recycling. A sudden change in conditions may not be accommodated. A mature ecosystem has many diverse species and complex food webs; roots are well established, and nutrient recycling is efficient. This makes it more resilient.

FITTED TO THE ENVIRONMENT

L IKE their habitats, plants and animals undergo continuous change – not only through their individual lives, but also in the overall characteristics of their species. Few species that have survived into the present have remained as they were millions or even only thousands of years ago: the modern horse, for example, is six times larger than its 58-million-year-old ancestor. Such changes in anatomy, physiology and behavior are necessary to allow living things to keep pace with any changes in their environment. If conditions change and individual organisms cannot adapt, they die; and eventually their whole species becomes extinct. Thus, the species that now live on Earth have all adapted successfully to the conditions they live in.

Some of the changes they undergo may be temporary, as in responses to changing seasons; others have resulted in the ability to tolerate a stable environment that most other organisms find unsuitable.

A good example is the ways in which camels have adapted to the extreme heat and the low availability of water in their desert habitats. Camels store moisture in the folded lining of their guts, and reduce water loss by producing concentrated urine and dry feces. Their body

KEYWORDS

ADAPTATION
ANNUAL
BEHAVIOR
CACTI
DECIDUOUS
DORMANCY
EXTINCTION
EVERGREEN
EVOLUTION
FLOWER
HIBERNATION
XEROPHYTE

▷ The Arctic hare responds to seasonal changes by changing color to blend in with its background. The onset of shorter days in autumn is a signal for its coat to change from brown to white. By the time the winter snows arrive, it is perfectly camouflaged. In spring, as the snow melts, the coat reverts to brown. This enables the animal to remain camouflaged from prey all through the year.

△ Chameleons are known for their ability to change color almost instantly – it takes less than two minutes from start to finish. Camouflage is their main purpose, usually for hunting; they are skilled predators of insects. They also change color to regulate body temperature, and when courting or competing for a mate.

temperature rises slowly with the heat during the day and drops quickly at night, eliminating the need for sweating. Fat in their humps can break down to provide water. The hump also retains heat, allowing the rest of the body to radiate it. Camels can survive 10 months without drinking; when the opportunity arises, they can replenish their store by drinking 100 liters (22 gallons) in 10 minutes.

Plants also show adaptations to dry conditions – both seasonal and permanent. Desert plants are the most obvious example, but other adaptations are found outside deserts. In the African savannas and in the Mediterranean, some deciduous (broadleaf) trees and shrubs shed their leaves for the dry season to conserve water. The leaves of evergreen trees in cold climates are thin and wax-coated to retain moisture during the long dry winter. (The snow that falls is of no use to plants as long as it remains frozen.) Other plants, particularly annuals (which grow anew and die each year), survive dry weather by having a short life cycle that is triggered by the onset of rain. They quickly flower and produce a large number of seeds which remain in the ground until the next rains. In hot climates small plants grow in the cracks of rocks where it is shady and relatively cool; or they flower in the spring to avoid the heat of summer.

Many animals and plants show adaptations to reduce the danger of being eaten. Grassland plants from the South American pampas to the British Isles grow from the base so that they can continue to grow when their tops have been eaten by grazing animals. Ground-level plants such as wild thyme and trefoil (a member of the pea family) form mats with their roots, anchoring themselves to each other or to the ground and making it difficult for animals to tear away their tissues. Prickly plants such as thistles are armed against feeding animals, which cut their paws or mouths on the plant. Among animals, camouflage allows an animal to blend with its background – an adaptation useful to both predators and prey. Bright colors may signify that an animal is poisonous to eat. This is seen in some species of frogs, butterflies and salamanders.

Slight differences may occur between similar species as a result of adaptation. Among elephants, for instance, the size of their ears differs with the climate in which they live. African elephants live in hot open country and have huge ears to increase the surface area of their skin, thereby losing heat and keeping cool. Asian elephants, in contrast, live in cooler forests and have smaller ears.

▽ A plant growing where the wind blows from only one direction shows uneven growth, with most occurring on the sheltered side away from the wind. An identical plant growing in a sheltered spot has an upright stance with an even distribution of leaves.

■ Desert plants burst into flower after rainfall ABOVE. They complete their life cycle while there is water available. The cactus RIGHT is the perfect desert plant. Its spiny leaves minimize surface area and the loss of water through transpiration. The stem is usually covered in a thick waxy layer to reduce water loss. A few species have white hair to reflect the light and to trap a layer of air around the plant, reducing the rate at which transpiration can occur.

Flower bud

Spines reduce water loss and deter grazers

Waxy stem retains water

Wide-ranging roots maximize water collection

ADAPTATION AND EVOLUTION

SOME individuals of a species are more successful than others in adapting to their environment – they may be faster, and so able to avoid predators; or taller, so that they can reach higher up on a tree, and still feed when all the lower leaves have been eaten. The individuals best adapted to the prevailing conditions are the ones who survive. They tend to produce offspring with similar characteristics. If environmental conditions change, a different set of characteristics is required for survival. In this way, the species gradually changes – the process of evolution.

Isolation of a species is an important factor in evolution. When individuals of the same species become separated – for example, by a geographical barrier such as a river, ocean or mountain range – they adapt to their own environments. If these are different, they may evolve along different paths. First they are still be able to interbreed; but, if the barrier remains in place, they eventually evolve into different species, which cannot interbreed. Then, if the barrier came down, they would remain unable to interbreed.

Trees at the top of a mountain experience different conditions than trees of the same species growing in a valley. As a result, the two groups of trees may show different growth patterns. They may become so isolated that interbreeding is no longer possible. This is an example of gradual adaptation leading to evolution.

The ocean acts as a considerable barrier to many plants and animals. The unique species of birds and reptiles of the Galapagos Islands off the coast of South America are a classic example of how isolation has assisted evolution. Zoologists believe that a few individuals of various South American species made the crossing from the mainland to the islands. There, they found uninhabited volcanic islands and they filled the available niches. They had little competition and, because of the numbers of different niches available, they began to specialize by feeding on different foods. Finches, often referred to as Darwin's finches, colonized all the islands and are now the dominant species of bird on land. As a result of lack of competition, the finches of the Galapagos evolved further into 13 different species. All are very different from the single species of finch found on the mainland.

KEYWORDS

ADAPTATION
BACTERIA
EVOLUTION
GENE
ISOLATION
MUTATION
NICHE
RESOURCE
TOLERANCE

▷ Dolphins, as members of the whale family, evolved from land-dwelling mammals. Their adaptations to aquatic life include a streamlined body with a tail for propulsion, and a blowhole at the top of the head for breathing when the dolphin comes to the surface of the water – which may be only once an hour. Spotted dolphins, shown here, are found in the tropical Atlantic. Their spots offer camouflage in sun-speckled shallow water, where dolphins feed.

▷ Iguana species are found from tropical rainforests to cold mountain peaks. The Galapagos land iguana has adapted to the hot, dry conditions of some of the islands. It feeds on the few plants that can be found, including the cactus.

◁ The Galapagos Islands contain the world's only marine iguanas. Under threat, they head for land – because, until humans arrived, this species had no experience of predators on land. They coexist with land iguanas by occupying different niches. The marine iguana is found on rocks by the coast and feeds on algae under water. During the morning it warms up in the sun and then searches for food in the water. It returns to the rocks to warm up again. If it gets too hot, the iguana cools down in the water.

The same process occurred for all the other animals colonizing the islands, resulting in a unique assortment of creatures. Giant land tortoises are another example; eight subspecies are each unique to a particular island, and can be identified by their different shells.

The evolution of the Galapagos species took many thousands of years, but evolution can happen on a much shorter timescale. One example of this is the development of species of rat immune to the rat poison warfarin. Warfarin was very successful when it was first used, but a very small percentage of the rat population had a gene that made them resistant to the poison. These rats survived and reproduced, passing their genes on to their offspring. Now there are many populations of rats that are immune to warfarin.

In the same way, resistance to certain antibiotics builds up very quickly in the short-lived, fast-breeding bacterial population. Frequent use of broad-spectrum antibiotics led to the survival of a few strains of bacteria which had resistant genes; these bacteria reproduced and passed on their resistant genes to new generations.

Genetic change has also been seen in some grass species that have been found growing at mining sites on spoil heaps with a high concentration of heavy metals such as lead, nickel, cadmium, mercury, copper, silver and zinc. Normally these metals would kill any grasses, but the new grass strains carry genes that give them immunity and enable them to survive.

PEPPERED MOTHS AND AIR POLLUTION

The peppered moth (*Biston betularia*), which lives on the bark of trees, is an example of evolution on a relatively short timescale. Before the Industrial Revolution, most of the moths were speckled **1**, and the rarer black moths experienced heavier predation by birds. During the Industrial Revolution, the bark of trees in industrial areas of Britain became blackened with soot. Darker and black peppered moths were protected by camouflage **2** and **3**. They became dominant in industrial areas, whereas the natural speckled moths remained more numerous in rural areas.

During the 1950s and 1960s, anti-pollution measures improved the air quality in Britain. The amount of soot in the air declined, and trees gradually regained their normal colors. The advantage then shifted back to the speckled moths.

1 *Biston betularia*

2 *Biston insularia*

3 *Biston carbonaria*

POPULATION *Studies*

5

POPULATION – the number of living organisms in a given area – and the environment are intimately linked. Understanding how the numbers fluctuate enables ecologists to identify the flow of energy through the ecosystem and the effect of one population on others. The study of the numbers of organisms in a population, and how the numbers change and respond to the environment, is called population ecology. The study of the periodic rises and falls in size of a population is called population dynamics.

The size of a population depends on many different factors. Competition for food and space, the incidence of disease, predation and climate are among the most important. A population increases in number when the number of births exceeds the number of deaths, and decreases if deaths exceed births. The size may also change if individuals leave or enter the populations from other populations.

Nature demands a balance in the size of populations. If they are too small, the species may be unable to reproduce in sufficient numbers, eventually becoming extinct; if they are too large, environmental resistance may limit the numbers by killing off individuals through famine, disease or other means. Humans have devised an astonishing range of methods to attempt to control the environment, but even we remain subject to these natural laws.

Firebugs cluster together at close quarters. Every species has a maximum population that its environment can support indefinitely. This is called the carrying capacity. Unfavorable conditions such as limited resources, competition and disease usually mean that populations' growth is restricted and they do not achieve their carrying capacity. Occasionally, however, conditions allow a "boom", and the population may grow larger than the carrying capacity – or, more commonly, environmental change causes the carrying capacity to be decreased. A population "crash" (sharp decline) results.

SURVIVING TO OLD AGE

REPLENISHING the numbers of a population is crucial to its stability and long-term survival. For a population to remain stable, on average every pair of individuals must survive to reproductive age and produce two offspring.

Population size is determined by the number of births and deaths. However, it is not only the absolute number of deaths which is important, but also the lifespan, particularly if a significant number of animals die before reaching reproductive age.

All species have a finite lifespan: there is an age by which all the individuals will have died. In humans, the average age span in developed countries is approximately 75 years. Before this, mortality is relatively low, but after this age, the death rate is very high, and only a few individuals live to a greater age; the limit for humans is about 110. In other species, deaths may be more or less evenly distributed across all age groups. This may be due to predation by larger animals, or to vulnerability to harsh weather or food shortages. In still others, infant mortality may be very high, so that few individuals ever attain the maximum possible lifespan.

Animals have evolved different strategies to ensure successful reproduction, given the limits of their lifespan and environment. Most fish have a high infant mortality rate; few survive into adulthood. Those females that do survive to breed tend to lay several thousand eggs all at once. They do not care for the

▷ No species gives more intensive care to its young than humans do. This design of nature has been supplemented by modern medicine, which – where available – further improves the survivorship rate. The weakest newborns of most species simply do not survive; but incubators nurture frail human young until they can thrive.

▽ In long-lived species such as elephants, humans and gorillas, infant mortality is low, and offspring tend to be fewer but better equipped for survival. The overall survival rate is therefore high until old age, when it suddenly begins drops off.

▷ Mountain goats lead a risky life. They spend much of their day climbing steep mountain slopes in search of grazing, often leaping from cliff to cliff. Although they are very nimble animals, they frequently have accidents – particularly the old, who have lost coordination or eyesight, and the very young, who are not yet experienced jumpers. For this reason there is a steady loss of animals from the population.

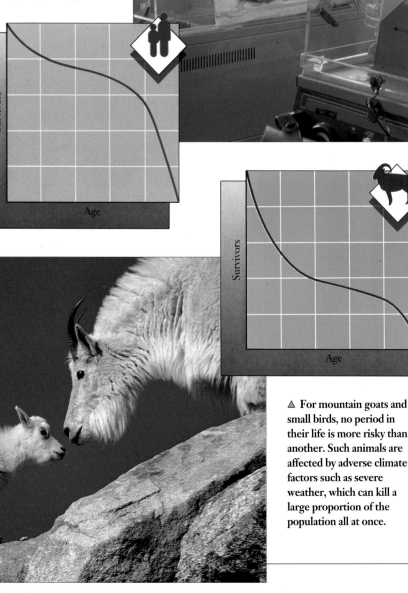

△ For mountain goats and small birds, no period in their life is more risky than another. Such animals are affected by adverse climate factors such as severe weather, which can kill a large proportion of the population all at once.

▽ Turtles are aquatic animals but the females have to lay their eggs on land. They go ashore at low tide and excavate large holes on the beach in which to lay their eggs. Over the next few hours they lay several hundred eggs. These are covered with sand, and the female returns to the sea. The eggs hatch a few weeks later, usually coinciding with hide tide. The turtles emerge from the sand and make their first journey to the sea. The huge numbers of hatchlings appearing en masse attract predators such as birds, which kill many of the young turtles. Only a few hatchlings successfully complete this journey. Those that do survive have a good chance of living to old age.

eggs, but leave the young to fend for themselves. The unprotected eggs attract predators, but, because the number of eggs is so enormous, a few do survive. If fish were to lay fewer eggs more frequently, predators would probably eat them all. Some species of fish have specific spawning seasons when many individual fish congregate and millions of eggs are laid together.

Large long-lived animals, especially mammals, are able to afford a different strategy. They produce a small number of offspring but each is given intensive care – feeding, protection and "lessons" in survival – throughout its youth. By actively caring for the offspring until it is old enough to survive on its own, the parents increase the young's chances of survival. This is particularly true of humans, who have to care for their young for at least 10 years – but the survival rate in this case is no longer uniform. Instead, it has come to be heavily influenced by factors outside nature. Wealthy countries with good medical care have low infant mortality rates, pushing up the average life expectancy of their population. Infant mortality in poor countries can be 3000 percent higher, which drastically reduces average life expectancy.

▽ Many plant species and animals, such as turtles and fish, produce thousands of seeds or eggs. Only a small proportion will ever develop further. Those who survive the first critical period of their lives have a good chance of living long enough to reproduce.

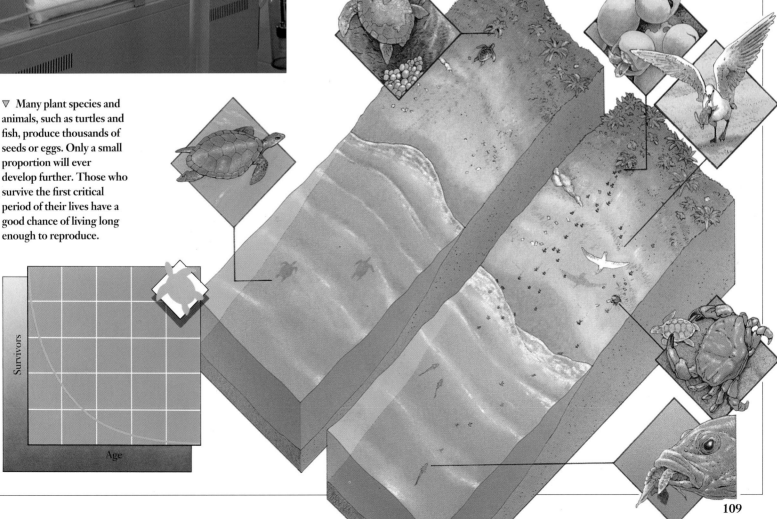

POPULATION CURVES

When an area is first colonized by a few individuals, there is plenty of food and space. They begin to breed, and the numbers slowly increase. Initially, the increase is limited by the number of females in the population and the time it takes to produce young. After a while, there are more females and the population reaches its maximum rate of increase, the exponential phase. The population increases until it reaches the maximum size that the area can support: its carrying capacity. Once this point is reached, the number of deaths equals the number of births and there is no further increase. As the population nears its carrying capacity, environmental resistance builds up in the form of disease or famine, for instance, and causes the rate of increase to slow. Eventually the population is in equilibrium with the environment.

A different pattern of population change is "boom and bust". Here, the numbers increase exponentially until environmental resistance suddenly causes the population to crash. The cause may be climatic change, or possibly interference by humans. A classic example is the lemming, whose population goes in four-yearly cycles. The colonies grow quickly until they reach a maximum density of 330 creatures per hectare, then the population crashes to 50 per hectare. The crash is linked to food supply, the spread of disease and social problems resulting from high density. Norwegian lemmings may undergo mass emigration in a search for new food sources.

Other species may be important to the cycle of growth of a population. There is a close relationship between the population sizes of prey and predator in any particular environment. For example, the number of snowshoe hares in Canada is influenced by the number of lynxes. The hares

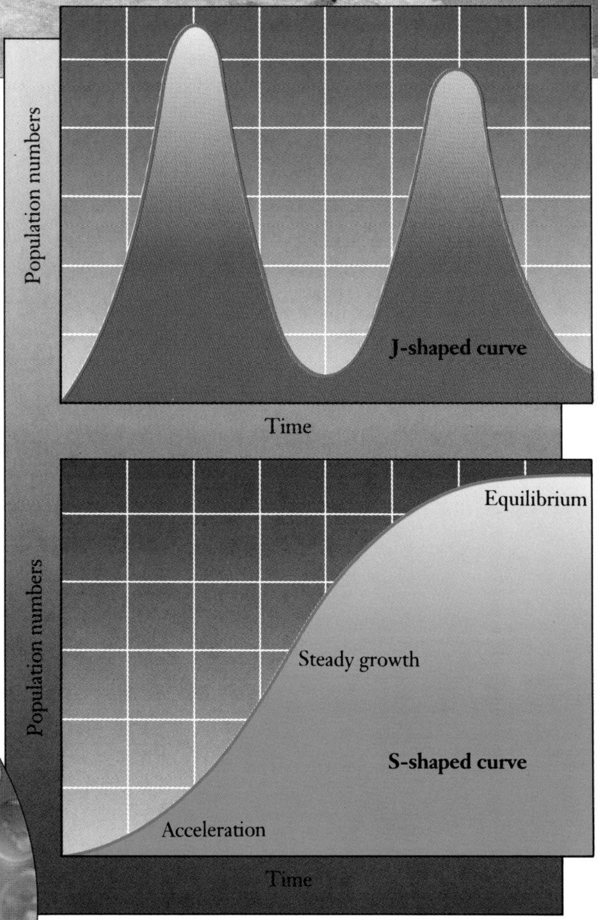

KEYWORDS

CARRYING CAPACITY
EXPONENTIAL
POPULATION
PREDATOR
PREY

△ Any growth in a population of prey animals, such as hares, results in more predators, such as bobcats. When the prey animals become fewer in number, the number of predators also falls; the two species thus show J-shaped population curves, slightly out of phase due to adjustment times. The bobcat helps to control the numbers of hares at a sustainable level.

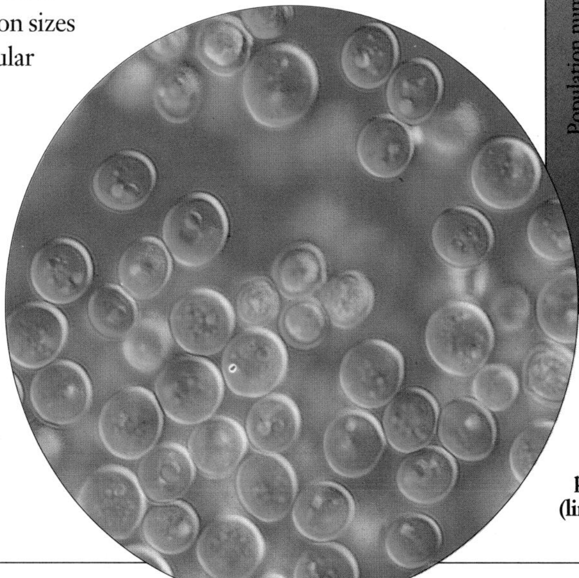

▷ A population of bacteria or yeast cells, shown here in a culture dish, grows slowly at first, then exponentially until it has filled the dish, doubling in size every few hours. When the dish is full, the increase slows and stops.

△ Two contrasting patterns of population expansion can be plotted on a graph. The pattern of steady increase (limited by environmental factors) is called the S-shaped or sigmoid curve ABOVE; the pattern of rapid increase followed by rapid decline is called the J-shaped curve TOP.

form 80 to 90 percent of the diet of the lynxes. An increase in the number of snowshoe hares results in a rise in the number of lynxes, and vice versa.

There are many techniques for monitoring the size of a population. The simplest is direct observation. This is possible with large animals which are easy to observe and count (elephant populations may even be monitored by satellite), and large plants such as trees.

For smaller animals, and species that are very mobile, other methods are used. Flying insects can be caught in a sweep net, and moths in a lighted night trap. Small mammals can be caught in traps and marked for further observation. Very small animals, such as beetles and wood lice, can be marked with nontoxic paint without harming them.

Population size can be estimated from samples using various statistical calculations. In the technique of capture–recapture, the animals are caught, the numbers noted and all the animals tagged or marked. Several days later, the trapping procedure is repeated and the number of animals bearing the mark is noted. An estimate of the population size can be made from the proportion of tagged to untagged animals.

▽ There are many interrelated factors that affect the populations of species living on savanna grasslands. One of the most fundamental is water availability. If water supply rises, this will increase the populations of large grazers such as wildebeest and buffaloes. These will drink more water, and tend to limit the rise in supply. Meanwhile, healthier animals mean less food for scavengers such as vultures, but more for predators such as lions, which hunt the young members of a grazing herd. Conversely, the growth in grasses that results from the increased water supply may stifle the growth of other small plants which creatures such as antelope eat; this will, in turn, affect the population of the antelope's predator, such as the cheetah. If, however, the rainfall fails, the likelihood of fires increases. Grassland fires result in the germination of acacia seeds, and the growth of new trees. These are eventually browsed by giraffes, whose population will therefore rise. The rise in the giraffe population will in time reduce the acacia coverage in the area, and the grass will return.

1 Savanna waterhole
2 Wildebeest
3 Vulture
4 Buffalo
5 Lioness
6 Grasses
7 Antelope
8 Cheetah
9 Fires
10 Acacias
11 Giraffe
12 Acacias

CONTROLLING POPULATION

THE ideal size of a population is close to the carrying capacity of its environment, but few populations remain static for long. If a population increases beyond the carrying capacity and becomes too large for the system that supports it, various environmental factors act to control its size. Some of these are purely physical factors, some relate to other species, and some are intrinsic to the species itself.

Among the physical factors, the availability of light is crucial for plants; a seedling is unable to grow under a dense canopy of leaves belonging to mature trees, but the death of a tree causes a gap in the canopy and allows a replacement to grow. Lack of oxygen can similarly restrict population size, especially of animals in small bodies of water such as ponds and fish tanks; the shortage of oxygen produced by overcrowding may cause the weaker individuals to suffocate. Climate change, whether drastic and sudden or more gradual, also affects the population size and composition.

The availability of food, and competition for an existing food supply, is a factor that relates to the populations of other species in the environment. Food shortage causes competition between individuals and the weaker specimens will inevitably starve.

When a species exceeds the carrying capacity of its environment, there may be emigration in search of new resources – lemmings provide an example of such behavior. There may alternatively be a crash, or catastrophic decline, in the size of a population; or factors within the population itself may automatically limit its growth. Thus, for example, disease spreads more rapidly in a large, closely-packed population than in a small, widely-spaced one:

in a densely planted field of wheat, fungal diseases such as rust and mildew spread rapidly. In many populations, overcrowding and other factors lead to stress in animals, and the birth rate naturally falls. For example, female rats do not breed in overcrowded conditions. The biological processes that control their fertility simply switch it off temporarily.

Most of these factors tend to favor the survival of the stronger specimens of a population, and they may result in shifts in the gene pool toward new adaptations. When a virulent disease strikes a population, it affects primarily the young, old and weak; but if part of the population carries a genetic adaptation that renders it immune to that disease, its members flourish, passing on their genes to the next generation.

When limiting factors – for example, food supply – depend on the size of the population, they are known as density-dependent factors: the greater the population, the greater the limiting effect. However, not all factors work like this. Climatic change and fire are examples of controlling factors that do not depend on population size: a series of frosts in a severe winter is equally likely to kill any plant or animal, regardless of the density of the population of which it forms part. These are examples of density-independent factors.

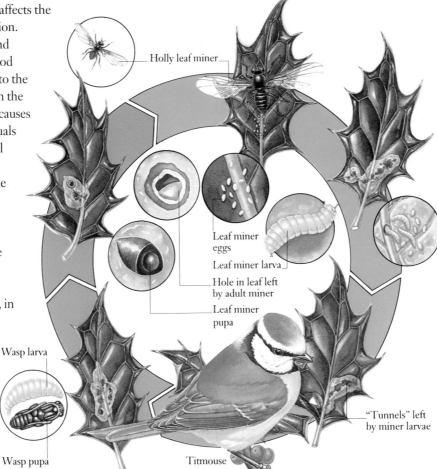

Holly leaf miner

Leaf miner eggs

Leaf miner larva

Hole in leaf left by adult miner

Leaf miner pupa

Wasp larva

Wasp pupa

"Tunnels" left by miner larvae

Titmouse

▷ Small birds are affected by many factors such as food supply and weather. A titmouse establishing a nest and rearing young early in the season has a greater chance of rearing more offspring successfully. Later

◁ The holly leaf miner is a small insect that lives on a particular holly tree. It lays its eggs on young leaves in summer. When the larvae emerge, they feed between the layers of leaf epidermis, forming blotch mines. Some larvae are killed by disease during the autumn. During the winter titmice may eat the larvae. In spring, a fixed proportion of the miner pupas are attacked by parasitic wasps, and wasp larvae develop within them. The surviving miners pupate in spring.

Many climatic factors act on a population independently of the size of the population. Severe winter weather can kill a large proportion of the population of creatures such as deer. The cold will attack all members of the population equally. Drastic environmental changes such as sudden floods may similarly wipe out large or small populations, regardless of the strength of the individual specimens, or the carrying capacity of the original environment.

in the season the decline in the food supply and colder weather may mean that a smaller proportion of the offspring are raised successfully. To survive a harsh winter, birds must seek new sources of food.

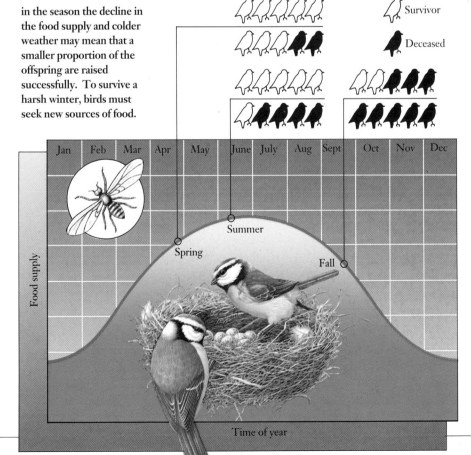

Sometimes, both density-dependent and density-independent factors act together to control a population. The holly leaf miner population on any individual tree, for example, is regulated by a predatory parasitic wasp, whose numbers are in turn dependent on the population density of fly larvae. However, the leaf miner larva has other predators: birds such as titmice feed on them too. The titmouse is acting as a density-independent factor on the miner population; during a cold winter it may eat a large percentage of the larvae present in a tree and the population can sometimes be wiped out.

With so many larger ecosystems affected in some way by human activity, the natural controls are no longer always enough to keep some populations within the carrying capacity of their environment. It then becomes necessary for humans to intervene, for example by culling – identifying those animals that can compete for food yet do not contribute to the social or genetic wellbeing of the community, and killing them. Culling is particularly important for larger animals that graze on lands that have been restricted by agriculture, such as deer on moors, or elephants in the national parks of southern Africa.

COLONIZATION STRATEGIES

W HEN a habitat is severely disrupted and the existing species are removed, perhaps through fire or by land clearance, the area effectively becomes a new habitat to be colonized afresh by plants and animals. Different species adopt different strategies when colonizing a new habitat, and their populations increase at varying rates.

The first species to colonize a new habitat are the kind of plants often referred to as weeds. These reproduce quickly and spread before other species have a chance to compete. They release large amounts of light seed that can be carried far away, often on the wind. Weeds put relatively more effort into reproduction than into structural growth, and are usually annual plants, living for one growing season and then dying. Some animals, such as flies, also adopt a similar strategy of quick reproduction, rapid growth and short lifespans: they too are among the first colonizers of new habitats.

Rapidly-increasing species are collectively referred to as "*r*-species". The arrival of such species of plant is often accompanied by animals that depend on that plant for food. For example, a field colonized by thistles attracts large flocks of goldfinches.

Other species grow more slowly and have a lower rate of increase. They are long-lived and devote far more of their resources to structural growth (in the case of plants, putting down deep roots and making woody trunks and branches). They tend to be found in stable habitats where the processes of colonization and change have been completed. These are known as "*K*-species" (*K* represents the carrying capacity). Larger plants such as trees and shrubs are examples of *K*-species. They are slow to arrive in a new habitat, but when they do, they compete successfully with the *r*-species for the resources. Tall, leafy trees deprive weeds of light and their root systems allow them to obtain more nutrients. *K*-species plants produce fewer seeds but do so over a longer period of time, and each seed tends to be relatively large, with more food resources for the embryo within it.

Should the habitat change again, *K*-species are less able than *r*-species to adapt to the new environment. This means that, as human impact on the environment grows, these species can become threatened. Animal examples of *K*-species include large mammals such as apes and elephants, birds such as the condor and albatross, and some large tropical butterflies.

KEYWORDS

CLIMAX COMMUNITY
COLONIZATION
HABITAT
K-SPECIES
R-SPECIES
WEED

If an area of temperate forest is destroyed by fire, a new habitat is created, which is recolonized by different species. The first species to arrive may be annuals such as rosebay willowherb, or perennial weeds such as nettles or dandelions, which produce a large number of wind-blown seeds. Animals such as the hawksmoth caterpillar, which lives on the rosebay willowherb, or the goldfinch which eats little other than the seeds of thistles, are also found at an early stage. Other plants, which arrive later, include grasses, docks and brambles, together with birds such as bullfinches that feed off seeds and berries. At this stage, a few tree seeds, perhaps brought by birds or squirrels, may germinate and begin to grow. If the seedling successfully establishes itself, its strong root system gradually takes up more and more of the available nutrients; by the time it has grown into a mature tree, such as the oak shown here, it dominates the area and few of the original colonists survive in its shade. The mature wood supports many animals, including jays and deer.

114

△ Certain species of plant
have adapted to factors
such as fire. The banksia is
found in grasslands where
fire is a major factor. The
plant produces cones
containing seeds that
germinate after a fire,
making use of the supply of
nutrients in the ash. Most
competing species will have
been destroyed in the fire,
allowing the banksia to
dominate the habitat.

OUT OF CONTROL

Approximately 20,000 years ago, humans started to use complex tools, improving their ability to hunt and find food. The result was a population boom. The beginnings of agriculture about 10,000 years ago caused further increases in numbers. The latest population explosion coincided with the start of the Industrial Revolution in the late 1700s and has continued ever since. Due to improvements in agriculture, medicine and industry, the Earth's population has risen exponentially; during the 1960s and 1970s the population curve went almost straight up; in 1990 the population of the Earth was over 5 billion, and it is projected to double by the year 2020. Natural life expectancy has increased, to an average of 75 years in the developed world. Lower death rates in almost all countries, combined with higher birth rates in some parts of the world, have caused this exponential curve.

Much of the industrial world, including western Europe, North America and Japan, has a stable population with births equalling deaths. They have a high standard of living with good medical facilities, education and available contraceptives. In under-developed countries, by contrast, populations continue to increase. More than 75 percent of the population in Kenya is under 20 years of age, which guarantees a vast future increase as this generation matures and has

▷ Up until the past century, population increases were small, and factors such as disease, famine and war controlled the population size. Now the increase has picked up speed. In 1900 there were 1.6 billion people in the world, and people in developing countries outnumbered those in developed countries by about 2 to 1. By 1990 there were 5 billion, and the ratio was 5 to 1. This pattern is predicted to continue. By the year 2100, the world's population could possibly be as high as 28 billion.

▷ With ease of movement around the world, the differences between human races are decreasing. An event such as the Olympics emphasizes that humans are all members of a single species.

South Asia

East Asia

Africa

S. America

N. America

USSR

Europe

1900	1925	1950	1990	2020
1.6 billion	2 billion	2.4 billion	5 billion	9 billion

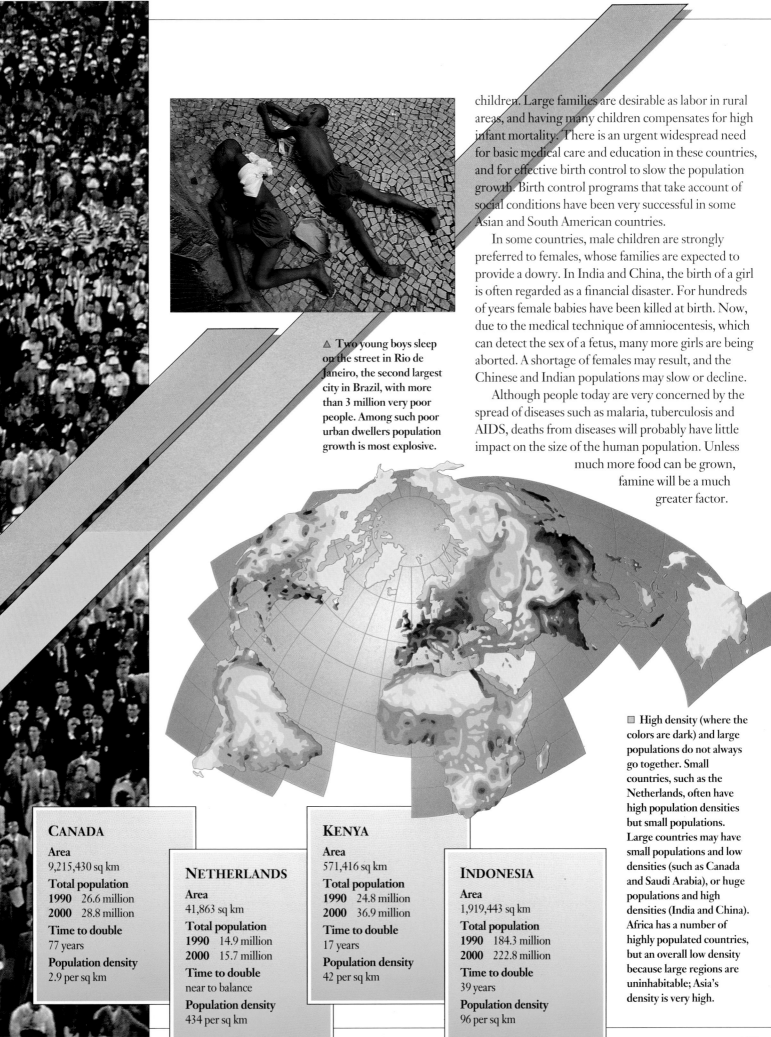

children. Large families are desirable as labor in rural areas, and having many children compensates for high infant mortality. There is an urgent widespread need for basic medical care and education in these countries, and for effective birth control to slow the population growth. Birth control programs that take account of social conditions have been very successful in some Asian and South American countries.

In some countries, male children are strongly preferred to females, whose families are expected to provide a dowry. In India and China, the birth of a girl is often regarded as a financial disaster. For hundreds of years female babies have been killed at birth. Now, due to the medical technique of amniocentesis, which can detect the sex of a fetus, many more girls are being aborted. A shortage of females may result, and the Chinese and Indian populations may slow or decline.

Although people today are very concerned by the spread of diseases such as malaria, tuberculosis and AIDS, deaths from diseases will probably have little impact on the size of the human population. Unless much more food can be grown, famine will be a much greater factor.

△ Two young boys sleep on the street in Rio de Janeiro, the second largest city in Brazil, with more than 3 million very poor people. Among such poor urban dwellers population growth is most explosive.

◻ High density (where the colors are dark) and large populations do not always go together. Small countries, such as the Netherlands, often have high population densities but small populations. Large countries may have small populations and low densities (such as Canada and Saudi Arabia), or huge populations and high densities (India and China). Africa has a number of highly populated countries, but an overall low density because large regions are uninhabitable; Asia's density is very high.

CANADA

Area
9,215,430 sq km

Total population
1990   26.6 million
2000   28.8 million

Time to double
77 years

Population density
2.9 per sq km

NETHERLANDS

Area
41,863 sq km

Total population
1990   14.9 million
2000   15.7 million

Time to double
near to balance

Population density
434 per sq km

KENYA

Area
571,416 sq km

Total population
1990   24.8 million
2000   36.9 million

Time to double
17 years

Population density
42 per sq km

INDONESIA

Area
1,919,443 sq km

Total population
1990   184.3 million
2000   222.8 million

Time to double
39 years

Population density
96 per sq km

6

THE GREEN
Revolution

FOR MOST OF THE HISTORY of the human race, people
lived by hunting wild animals and fishing for food, and
gathering edible plants. Very few people still live this way –
the Kalahari bushmen of Namibia among them.

About 9000 years ago, in the Middle East, people began to
capture antelopes, and gather the seeds of grasses to feed them.
These simple changes, which also occurred in northern China and
in Central and South America, were the beginnings of agriculture.
The first farmers became familiar with the life cycles of the plants
they gathered and began to grow their own. These were the grasses –
the forerunners of modern cereals.

Gradually, as a result of thousands of years of selection of the
best-producing plants by farmers, food crops improved. As seeds
were transported to different parts of the world and planted, they
were pollinated by wild species, occasionally creating improved
varieties. Modern crops are far removed from the ancestral species.

Animals changed in a similar way. Modern cattle are very
different from wild oxen. At first, farmers kept wild animals for meat
or milk, but soon began to domesticate them. The largest and most
docile were used for breeding. As a result, species such as cows, pigs
and goats grew larger, providing more milk and meat.

Rice seedlings in paddy-fields in China are still planted by hand. Few farmers have modern equipment, and the seedlings are so delicate that they require the most careful handling. The world's most populous nation, China has a long history of famine. However, since the 1960s it has increased its production of rice (one of its main crops) by more than 100 percent and almost eliminated malnutrition. Much of this improved food supply is needed to feed the populations of China's cities, which are growing rapidly in spite of strict controls on rural-to-urban migration.

FEEDING THE WORLD

As the human population rose, agriculture had to become far more intensive to support it. Much of the available land on Earth is now used to grow the 30 principal crops that feed the world; of these, wheat, rice, corn (maize) and potatoes are most important. Scientific research continuously seeks new ways of improving yields: by farming more land; by increasing the productivity of existing farmland through the use of fertilizers; by developing new crops with higher yields; and by decreasing the proportion of the crops that is lost to pests.

Since the 1940s there has been a dramatic increase in monoculture, the continued planting of the same crop on the same land for several years (instead of planting different crops from year to year as in traditional agriculture). Economies of scale can be applied to monoculture. Huge uniform fields allow the use of large, efficient machines such as combine harvesters. However, monoculture creates many problems, including nutrient depletion, pest buildup and soil erosion. To keep yields consistently high, the crops must be heavily fertilized and treated with pesticides.

Plant breeding plays an important role in modern agriculture. New varieties of crops are being developed that have greater yields and resistance to disease and pests. Hybrids and dwarf crops are examples. Triticale is produced by crossing wheat with rye. It has the same yield as wheat, but is more nutritious and has the winter hardiness of rye. It can also be grown on marginal soils. So far it is used only as an animal feed but, in the future, it may be eaten by humans. Rice, too, has undergone significant changes. The modern rice plant is semi-dwarf – bred to be shorter than normal, so that most of its nutrients are concentrated in the grains rather than in its non-edible structures. It also matures early. However, without the addition of fertilizers and a constant supply of water and pesticides, the crop fails. For many farmers in the poorer parts of the world, native varieties are better suited to the climate.

The introduction of such crops transformed world agriculture as the techniques spread from developed to developing countries. Between 1950 and 1988, in this "green revolution", world food production increased

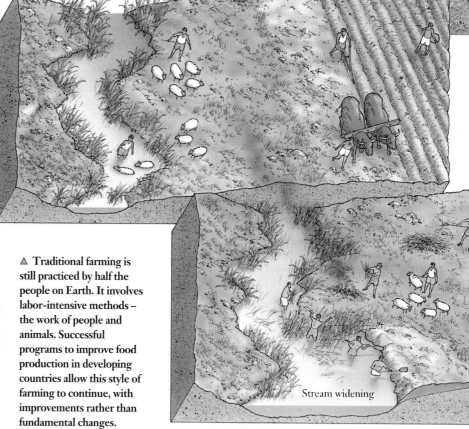

△ **Traditional farming is still practiced by half the people on Earth. It involves labor-intensive methods – the work of people and animals. Successful programs to improve food production in developing countries allow this style of farming to continue, with improvements rather than fundamental changes.**

Stream widening

◁ **In hydroponics, plants and flowers are grown in large containers with circulating nutrient-enriched water. The light is adjusted to stimulate growth. This technique is particularly useful in arid climates.**

▷ **Mechanized farming is more efficient on a large scale, but many poor farmers cannot afford equipment and must find other ways to improve their production. Irrigation supplies water to crops that require more moisture, and** allows farming on land that is otherwise too dry; drainage improves land that may be too wet. Finally, building roads allows the farmed produce to be transported to markets, whether local or on the other side of the world.

by 140 percent, and average prices food dropped by 25 percent. Much of this increase was achieved in developed countries and in Asia, especially in China. By contrast, in 43 developing countries (half of them in Africa), food production dropped 20 percent during this period, due to factors ranging from wars and drought to poor soil and lack of investment. Production also began to decline in developed countries from the late 1980s, due to soil erosion and pollution.

The issue of what people eat is becoming increasingly important. In 1982 a United Nations study in 117 developing countries showed that farming yields could, in theory, support a global population up to 32.4 billion, against the current level of about 6 billion – but only if everyone gave up eating meat and milk products, and if every available meter of land was used to grow crops (with none left for trees or wildlife). But if every country in the world was to have access to the same range and quantity of food that most developed countries now have, the maximum that could be supported would be only 8.5 billion.

△ As the world population increases, there is ever more pressure to find ways of cultivating land that is not ideally suited to agriculture. Terracing is one such technique. It has

been used in southeast Asia for 2000 years, allowing people to cultivate land that is otherwise unsuitable for farming because it is too hilly. Terracing also helps prevent erosion of the slopes. These terraces in the Philippines are now used as rice paddies. For the first part of the growing season the fields are kept under water; then, as the crop ripens, the fields are drained.

◁ Intensive farming increases meat production by keeping large numbers of animals in special barns. Diet, heat and light are designed to maximize growth, and the animals are allowed little space to move so they conserve energy. But growth is inhibited due to the stress of these unnatural conditions.

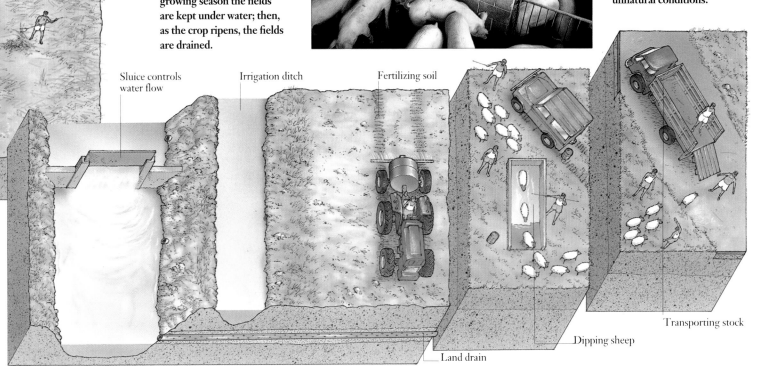

Sluice controls water flow

Irrigation ditch

Fertilizing soil

Transporting stock

Dipping sheep

Land drain

INTENSIVE FARMING

MODERN crops need high levels of nutrients to support their growth. The three most important mineral nutrients are nitrogen, phosphorus and potassium. Without them, plant growth is stunted, their leaves turn yellow and yields are poor. These minerals are the main ingredients of fertilizers. Some fertilizers are byproducts of organic materials, whereas others are derived from chemical processes or mining, or manufactured specifically for the purpose. They are applied by hand or by spraying from aircraft, dissolve rapidly and are absorbed by the soil or taken in by the roots of plants.

In a monoculture, the large number of similar plants grown in a small area attracts pests. To control them, farmers spray crops with pesticides. Insecticides are usually synthetically manufactured substances containing chlorine or phosphorus. They may be sprayed directly onto the pest, or applied to the crop to poison the pest that feeds on it. Fungicides are used to destroy fungi, which are a common cause of disease in plants. The ancient Romans are thought to have used a mixture of copper sulfate, quicklime and water to control fungi on grapevines, and sulfur has been widely used as a fungicide since the early 19th century.

Animals need a high input of nutrients, particularly protein, in order to grow quickly – one of the chief goals of intensive farming. Foods derived from animal tissue contain more protein than cereal-based feeds.

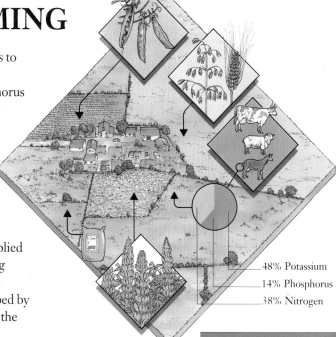

48% Potassium
14% Phosphorus
38% Nitrogen

◁ Traditional farming involved rotating the crops on a piece of land and then leaving the land fallow for one year to recover. A cereal crop might be followed by a root crop and then a member of the pea family such as peas, clover or beans, which increase the nitrogen content of the soil. The crops use different nutrients, so the soil is not depleted; and each has different pests, so one pest species does not become well established.

▷ A tractor and a combine harvester cut swathes across a field in France. The French farm industry did not become modernized until after 1945, when government incentives gradually began to reduce the workforce of farmers, and small holdings were consolidated into giant commercial farms. The use of fertilizers soared as this land was intensively cultivated. Cereal production has more than doubled since 1955. France is the sixth largest exporter of wheat and barley in the world.

▷ Arable weeds such as poppies, corn marigolds and corn cockles often colonize cleared land after a harvest, or at the edge of a crop. Weeds compete with crops for nutrients, water and light, so farmers spray the field with herbicides – chemicals that inhibit plant growth. Herbicides are designed to harm only broadleaf weeds, not the narrow-leaf crops they flourish among.

For this reason, many farmed animals – even those that are naturally vegetarian, such as cows – are fed on slaughterhouse waste. Such foods can cause problems. In the late 1980s many European cattle were infected with bovine spongiform encephalitis (BSE), a disease of the brain, that was transmitted from sheep infected with a disease called scrapie. Cattle infected with this disease had to be destroyed, and it is not yet known whether the disease could have been transmitted to people who ate the beef, or with what effects.

Research has shown that giving animals extra hormones can improve their growth. Some of these have the potential to be developed from artificial sources. The use of bovine growth hormone in dairy cows increases milk yield by 20 percent or more.

In the early 1990s the hormone was being tested for mass-market use in the United States, but it was banned in most European countries.

For the most part, intensive farming techniques have bypassed the subsistence-level farmers who are the majority in developing countries. They have little land, and cannot afford to buy equipment, fuel and agrochemicals, but continue to improve production by traditional labor-intensive methods. China has been particularly successful in this way. Other developing countries have attempted to switch to industrialized intensive farming methods, but with disastrous results, including huge debts to overseas suppliers. This has undermined their economies, making them vulnerable to fluctuating markets as consumers and producers.

▽ Sheep are susceptible to external parasites, particularly in the crowded conditions found in intensive livestock farming. Most sheep are dipped in pesticides once a year to control these parasites. Farmed animals are also vaccinated regularly and are often given large doses of antibiotics against the outbreak of disease.

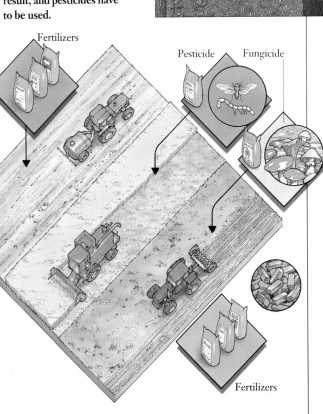

▽ Modern agriculture involves monoculture – the planting of the same crop in the same field for many years in a row. This method causes the soil to be depleted of the nutrients required by that crop, and artificial sources have to be applied in the form of fertilizers. Pest species also build up more easily as a result, and pesticides have to be used.

Fertilizers

Pesticide   Fungicide

Fertilizers

THE EFFECTS OF AGROCHEMICALS

THE use of fertilizers, pesticides and other agrochemicals worldwide increased tenfold from 1940 to 1988. Crop yields improved dramatically, and the incidence of diseases such as malaria decreased as pesticides killed the insects that carry them. However, the effect of so much chemical interference could not be contained within the farmed environment. Because of the scale on which agrochemicals are used worldwide, they are now present throughout the biosphere. Many of the older generation of agrochemicals break down very slowly, which allows them to linger in air, water and soil and to accumulate in the tissues of plants and animals. For this reason, even small amounts of these chemicals can cause longterm, widespread harm. More recently developed agrochemicals were designed to break down more quickly, but this makes them even more toxic to organisms that ingest them. It also means that they must be applied more frequently, so that they are always in the environment, like the slowly-degradable agrochemicals they were designed to replace.

Nitrates are among the most common ingredients of artificial fertilizers. Whether they occur naturally in soil or are supplied by fertilizers, nitrates are essential for providing nitrogen that is taken up by plant roots and allows plants to grow. However, an excess of nitrates may build up where water runs off the fertilized land. This results in explosive growth of algae in the water, which becomes choked and stagnant. Nitrates are harmful in other ways. High levels in drinking water are toxic to humans, particularly infants. They have been linked to fetal deformities, blood poisoning and gastrointestinal cancers. Nitrates in water often combine with pesticides (which include herbicides, fungicides and insecticides), making them even more lethal.

Pesticides are designed to kill, and they may take other species along with the pests they were intended to control. Frequently, the pesticide kills the natural predators of the pest. Because the local populations of predators are less able to control the pest by natural means, the balance of the local ecosystem is upset. Habitats may also be destroyed. For example, stretches

KEYWORDS

ACCUMULATION
AGROCHEMICAL
DDT
FERTILIZER
FOOD CHAIN
FUNGICIDE
INSECTICIDE
LETHAL DOSE
NITRATE
NITROGEN FIXATION
PERSISTENCE
PESTICIDE

△ Modern crops are sprayed with an assortment of agrochemicals to maximize their yields. Some farmers and scientists argue that agrochemicals are necessary to maintain soil nutrients and control pests from insects to weeds and fungi. However, pests build up resistance to specific chemicals, and ever greater amounts must be used – increasing the level of chemicals in the biosphere – or completely new chemicals must be introduced. This has led to 50,000 pesticide products in the United States market alone, applied at a rate of 454,000 tonnes every year.

▷ Tropical fruits are one of the major exports from developing countries to North America and Europe. They may be sprayed as they grow with agrochemicals that are regulated or banned in the countries that import them (although these same countries are often the producers of the agrochemicals). Only laboratory tests can tell whether they have been sprayed, and it is impossible to test all imports; the United States' Food and Drug Administration tests only about 12,000 samples a year, including domestic foods. Imported fruits are not the only ones that pose a risk. Ordinary crops such as lettuce, tomatoes, potatoes and grapes in developed countries are also typically heavily sprayed.

▷ Spraying crops with pesticides not only affects the environment; people who handle the chemicals, such as farm workers, may suffer from exposure if safety precautions are not followed. The World Health Organization estimates that 1 million people are poisoned by pesticides every year, and up to 20,000 may die. Many of these are farmers in developing countries.

◻ Many birds and small mammals are attracted to the seeds of cereal crops ABOVE. However, when they eat the seeds they take in residues of pesticides which have been sprayed on to the crop. The pesticides build up in their tissues. If they are then eaten by predatory animals or birds, the pesticides within their bodies are passed to the predator. In this way pesticides increase in concentration through the food chain. Thus, the worst effects are seen at the top, as in birds of prey such as this harrier.

of land along highways in industrialized countries were laid waste as pesticides were sprayed in an attempt to control weeds. Flowers and other plants also died, along with the animals they supported. When the spraying stopped, these ecosystems revived.

One of the most environmentally damaging agrochemicals is DDT, which was used extensively as an insecticide during the 1950 and 1960s. It does not break down rapidly in the environment, thereby acting for longer and requiring less spraying; and it is insoluble in water but soluble in fat, which means that it becomes concentrated in the fatty tissues of animals. Because of this, DDT was incorporated in food chains. Small amounts were present in water and taken up by the primary producers. These were eaten by herbivores, which ate large amounts of plants and accumulated DDT. As the herbivores were eaten by predators, the concentration of the chemicals built up. At the top of the food chain, birds of prey such as the peregrine falcon suffered metabolic disorders. Female birds laid eggs with a thin shell that broke under the weight of the mother. During times of starvation, particularly during cold winters, the fat was drawn on by the body, rapidly releasing large quantities of DDT into the bloodstream. In the severe winters of 1957 and 1963, millions of European birds died.

DDT has been banned in much of the world, but in some countries it is still used as a cost-effective control against the mosquitoes that cause malaria. Traces of DDT are found in the bodies of most people in the world, and even in penguins in the Antarctic, thousands of kilometers away from any source. It is not yet known what effect these small amounts will have over the course of a lifetime.

BIOLOGICAL ALTERNATIVES

Almost all agrochemicals have a damaging impact on ecosystems. Increasing environmental awareness and improved technology are combining to produce biological alternatives. Some of these may involve a return to traditional methods of cultivation, but with modern improvements, such as mechanized equipment or better irrigation. Others are developed in the laboratory and involve the invention of new crops, as in the high-yield varieties that were developed after the 1940s and led the "green revolution" across the developed world. New varieties of crops may be genetically modified to enable them to make their own pesticides. Genes that enable plants to fix their own nitrogen may also be inserted into cereal plants, reducing the need for artificial fertilizer. Genetic engineering is expected to add $20 billion annually to the value of food crops by the end of the century.

Biological methods of pest control are not only safer; they are often more economical – up to 10 times cheaper than pesticides. Pests have predators too, and it is possible to make use of them to reduce pest numbers dramatically. In California in the 1940s, $750,000 was spent to introduce a European beetle to control a particular weed that was killing cattle. It has saved ranchers over $100 million. Another example is the white fly, a common insect pest found on greenhouse plants. It can be controlled by introducing a small parasitic wasp called *Encarsia*. The wasp does not wipe out the white fly population, but reduces its numbers so that its effect is negligible. Animals also provide biological pest control. Ducks, chickens, geese and other birds feed on a variety of pests, from slugs and insects to weeds, though they may damage young and leafy plants, and their access to crops should be managed.

Intercropping is an equally effective technique that involves planting more than one crop in a field. Among mixed crops, pests cannot spread so easily. French and African marigolds keep nematode worms (eelworms) at bay by releasing sulfur compounds into the soil. These compounds affect the soil to a distance of a meter's diameter around the plant. This is why marigolds are often planted with potatoes.

These natural insecticides have tremendous potential. For example, in India, the neem tree produces a cocktail of chemicals that is effective against 100 different insect pests. The chemicals interfere with the insects' life cycle, preventing them from developing and so interfering with their reproduction. Plants of the chrysanthemum family produce pyrethrum, which is more harmful to insects than DDT, but is safe for mammals and plants. Plants also produce many natural chemicals to deter herbivorous animals. These are usually toxic to only one group of animals, and rarely persist in the environment.

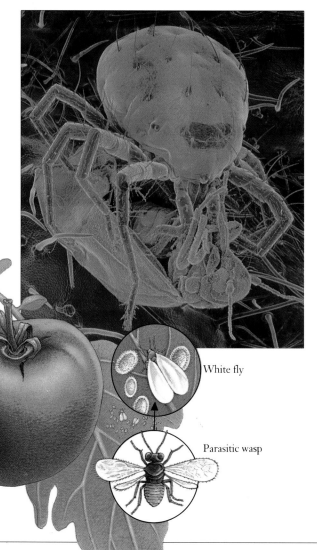

White fly

Parasitic wasp

■ Tomatoes BELOW LEFT are susceptible to infestation by white flies. If the white flies are sprayed with pesticides, these can be absorbed by the tissues of the tomato fruits, and poison people or animals who eat the tomatoes. A chance observation by a scientist at the Cambridge Botanic Garden in Great Britain led to biological control. The white flies are preyed on by a small parasitic wasp, which reduces the numbers of white flies. White flies can also be controlled by mites ABOVE LEFT. The mite attacks the white fly, gripping it and devouring it. Predators such as these are being investigated with the aim of introducing natural pest control in commercial greenhouses.

◁ A scientists tends a crop of genetically engineered corn (maize) in a California laboratory/greenhouse. Plants may be modified for pest control, for higher yields and for drought resistance – of urgent interest to farmers if global warming and desertification intensify as predicted.

▽ Strawberries are often planted in rows between cabbage plants to control pests that damage the cabbages. The strawberries release a natural chemical that is toxic to these pests but does not harm the cabbage crop. The crop of strawberries is a bonus for this method of pest control.

▽ The technique of intercropping was known to ancient Amerindians, who discovered that by growing the corn (maize) crop with a second crop of beans there would be better control of pests and increased yields. The beans used the main corn crop for support, while enriching the soil with nitrogen for the nitrogen-depleting corn. The corn yield was not affected, and at the end of the season there was the bonus of a crop of beans.

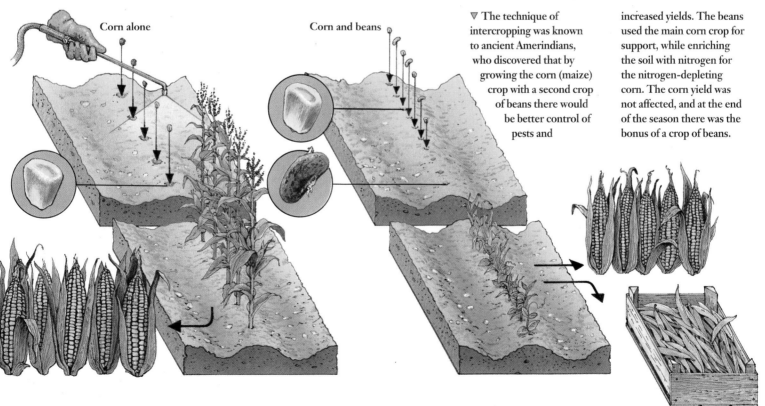

Corn alone

Corn and beans

OVEREXPLOITATION OF RESOURCES

THE area of the Earth that is suitable for growing food is limited. About 51 percent of the land is covered by ice, snow, desert and mountains. Most of the world's population lives on, and is fed by, another 21 percent of the land. As populations grow and food production expands, this land is under ever-increasing pressure. As much as 30 percent of farmland around the world is at risk from overuse. In developing countries, 65 percent of today's farmland may become unfit for farming in two decades.

Both largescale industrialized agriculture (as practiced in developed countries) and smaller-scale subsistence agriculture (in developing countries) contribute to this degradation. Erosion of the soil is one of the chief problems of overuse. More than 20 billion tonnes of topsoil are lost every year; China, India, the United States and much of the former Soviet Union are the worst affected. Industrial farming techniques such as monoculture (planting huge fields with one crop, year after year) deplete the soil of essential nutrients. Most crops lack strong roots to hold the soil together, and fields are plowed after each harvest and left bare between crops, allowing soil to blow away. If drought occurs, large regions can become "dust bowls", as in the central plains of United States in the mid-1930s. Industrialized agriculture also depends heavily on fossil fuels, making the food supply vulnerable to any scarcity.

In developing countries, rapidly-increasing populations are forcing the clearance of more land, and small farmers no longer leave plots uncultivated for several years to recover in between crops. Livestock overgrazing in arid and semi-arid regions is contributing to the destruction of plant cover and the spread of deserts at a rate of 6 million hectares a year.

Agriculture consumes huge quantities of water, ranging from about 41 percent of total use in the United States to 87 percent in China. The availability of water is crucial to growing food. About 40 percent of the world's people live in regions where there are frequent severe droughts. In Africa in the mid-1980s, drought killed tens of thousands of people through famine and disease, and forced the mass migration of 10 million. Competition for access to water is increasing. India is involved with disputes with Bangladesh and Pakistan over the Ganges and Indus river supplies, and access to the Jordan River is a major political issue in the Middle East. An increasing number of dams and reservoirs are being built around the world in an attempt to regulate water supplies. However, reservoirs become blocked with silt over time, and the flooding of land to form them disrupts ecosystems and displaces large numbers of people. Also, dams and reservoirs may actually decrease the water supply, because the rate of evaporation from a standing body of water is increased.

The demand for water for irrigating crops is expected to double as developing countries attempt to increase their food production. Conservation methods can make irrigation more efficient – for example, using a drip or trickle rather than a high-volume spray, and releasing the water underground near the roots, rather than on the surface. These and other techniques have allowed Israel to cut irrigation waste by 84 percent. Inefficient use of water is not only wasteful, it may also damage the environment. Without proper drainage, irrigation can result in soil becoming salinized or waterlogged, and fertilizers used on the land may run off into the water supply, polluting it.

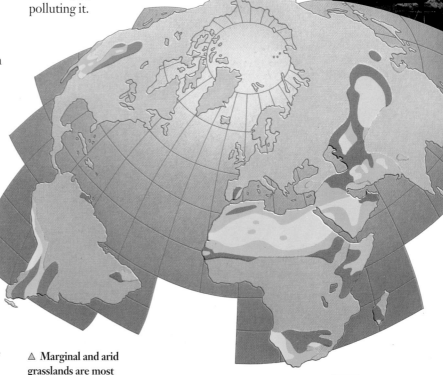

△ **Marginal and arid grasslands are most vulnerable to overgrazing and soil erosion, leading to desertification. Today many areas in the world are under threat of desertification, especially in** Africa. This reduces possible food production for 400 million people, and makes some 50 million unable to feed themselves.

Existing deserts

Areas of desertification

At risk of desertification

□ A towering sand storm LEFT approaches a village in Mauritania, at the edge of the Sahara desert in Africa. Such storms are common where land is so degraded that the soil simply blows away. This is one result of overgrazing, particularly by non-native species of cattle which are not adapted to local conditions. Such cattle have no natural immunity to disease and tend to overgraze the land. Farming wild native species such as eland ABOVE means that more animals can be supported on the same land without so much degradation.

▷ Whales are still hunted by traditional methods in the Faeroe Islands in the North Atlantic, but commercial whaling around the world has made many species nearly extinct. Other marine species are also being depleted through overfishing.

Heavy irrigation, even in areas where the water supply is apparently plentiful, leads to depletion of ground-water, one of the most important sources of fresh water for the planet, which is replenished only very slowly. Northern China, parts of India and Mexico, and the southwestern United States are all at risk of groundwater depletion. If depletion under ground is very severe, particularly near the coast, salt water may begin to leak in and contaminate the fresh water; this is a common problem around the Arabian Gulf.

7

THE
Human Factor

SINCE THE BEGINNING of the Industrial Revolution in the early 18th century, there has been a growing demand for fossil fuels and minerals, most of which are extracted from the ground. Mining, quarrying and road construction can wreak havoc on the natural environment on a large scale. Another consequences of widescale development is pollution, which now affects every part of the biosphere – air, water and land. Even the wastes of previous generations are proving toxic. Many of these pollutants become incorporated into the food chains, allowing the chemicals to pass from animal to animal.

Environmental problems are not the same throughout the world. Pollution in industrialized countries is mainly caused by burning fossil fuels as well as by inadequate disposal of industrial and domestic wastes. In poorer undeveloped countries, the rapidly increasing populations are creating ever more demands on the land. Largescale habitat loss results when forest and grassland are cultivated. Where sanitation is inadequate, rivers are often used as sewers. These waters are unfit for drinking and carry disease.

Ecologists have begun to remind people that the biosphere is in a delicate balance, and a change in one part often affects other parts. Once this balance has been upset, the damage is difficult to rectify.

A tangle of wire traps the skeleton of an animal on a beach – a jarring reminder of the effects of indiscriminate dumping of any kind, and the tendency of environmental damage to take unexpected as well as obvious forms. The oceans have long been used as dumping grounds for all kinds of wastes, from chemicals to plastic containers. Chemical waste may be invisible until marine life shows signs of poisoning, but plastics and other debris may wash up on shore, often attached to the bodies of the animals they have entangled and strangled. Other animals may mistake the debris for food, eat it, and die.

131

ATMOSPHERIC POLLUTION

Every day, tonnes of pollutants are poured into the atmosphere. There are many sources of air pollution but the primary ones are factories and vehicle emissions. The results are acid rain, smog, depletion of the ozone layer and global warming.

Acid rain is created by a cocktail of chemicals that includes sulfur dioxide (derived mainly from coal-burning power station emissions), nitrogen oxides and hydrocarbons. Rain itself is naturally slightly acidic, but in the presence of pollutants such as sulfur dioxide, it becomes even more so, forming sulfuric and nitric acid. Winds can carry the acid rainclouds thousands of kilometers from the original source, making the control of acid rain an international problem. Acid rain affects the normal acidity of the soil on which it falls, with naturally acidic soils being the most affected. Toxic elements such as aluminum and cadmium are leached from the acidified soil, and plants absorb them through their roots. Their leaves turn yellow and die; then whole branches die, especially those on the outside of a forest and at the top of the canopy. The loss of leaves results in reduced photosynthesis, making the plant less able to survive other stresses such as drought and frost, and leaving it more vulnerable to attack by disease. In some central European countries, such as Poland, it is estimated that three out of every five trees have been damaged by the combined effects of acid rain and drought. Acid rain also quickly affects the water in lakes, killing almost all their fish and invertebrates.

Another common form of air pollution is smog. This forms as a thick yellow layer above cities, particularly in those that are surrounded by mountains, such as Los Angeles, Mexico City and Athens. Smog is created by sunlight reacting with nitrogen oxides and hydrocarbons from vehicle emissions. When the layer of air immediately above the ground cools more rapidly than the air above it, an inversion layer forms, trapping the lower layer of air and pollutants close to the ground. Smog is particularly severe on sunny days with little wind, and several of its components, such as peroxyacetyl nitrate, can be especially dangerous to young children and people with asthma and other respiratory complaints.

Although ozone (O_3) close to the ground is poisonous to humans, high in the atmosphere it is vital to the survival of most forms of life. The layer of ozone in the upper atmosphere, 15–50 kilometers above the

KEYWORDS

ACID RAIN
GREENHOUSE EFFECT
INVERSION LAYER
OZONE HOLE
POLLUTION
SMOG

▷ Cities such as Los Angeles that lie in a natural hollow surrounded by mountains are susceptible to the buildup of a thick yellow layer of smog. Under certain weather conditions, the layer of air above the city is trapped and pollutants build up. The vehicle fumes together with factory emissions react in the presence of sunlight to create photochemical smog. Attempts to deal with the problem include the longterm reduction of vehicle emissions, and the banning of vehicles in the city centers on the worst days. Vulnerable people such as children, elderly people and asthma sufferers are advised to remain indoors during a smog.

surface of the Earth, absorbs ultraviolet (UV) radiation. Chemicals such as CFCs (chlorofluoro-carbons) react with the ozone, causing it to break down. Until the 1980s, CFCs were used extensively as propellants in aerosols and refrigerants. They are very stable, light molecules, and careless disposal has resulted in their entering the atmosphere. Once there, they remain for many decades. The result has been seen since 1982 as a severe thinning that appears in the spring in the ozone layer above Antarctica. The size of the "hole" is increasing – in 1987 it was as large as the United States – and the ozone layer is also beginning to thin over major industrial regions. The longevity of

◁ Vehicle exhausts release a collection of chemicals including carbon monoxide, sulfur dioxide, nitrogen oxides and hydrocarbons. Some gasoline also contains lead, which appears in the exhaust fumes and can have adverse effects on brain development in children. Throughout the world's cities, many people – such as cyclists, who have to breathe deeply while in close proximity to vehicle exhausts – have begun to wear masks to filter the air they breathe. Here, a Green Party protestor against the poor quality of city air in Rome emphasizes the point.

▷ **Chemicals such as CFCs move into the atmosphere where they release chlorine atoms. These break down the ozone molecules in the stratosphere into oxygen molecules. Incoming ultraviolet light is then no longer absorbed.**

Ultraviolet light
Oxygen atom

Oxygen molecule
Ozone
Chlorine atom
CFC

Sulfur dioxide
Nitrogen oxides
Hydrocarbons

Acid rain

Photochemical smog

■ **Sulfur dioxide, nitrogen oxides and hydrocarbons react in the presence of sunlight to form sulfuric acid and nitric acid. They dissolve and fall as acid rain ABOVE. The results can kill entire forests BELOW.**

▷ **Photochemical smog results from a complex series of reactions. They begin with the conversion, by means of strong sunlight, of atmospheric oxygen and nitric oxide from vehicle exhaust fumes,** to nitrogen dioxide, leaving a reactive free oxygen atom. This combines with hydrocarbons from the unburned fuel, to form a group of offensive and toxic chemicals that are the main components of smog.

the CFC molecule means that international restriction on their use will not reduce their concentration in the upper atmosphere for many decades.

Damage to the ozone layer will allow more UV radiation to penetrate the atmosphere, where it may damage living cells and cause eye problems and skin cancers. It may also damage phytoplankton. However, UV radiation may not be as harmful as first thought. Some experiments have shown that photosynthetic algae can survive exposure to UV radiation.

CFCs, like carbon dioxide, are also greenhouse gases and their presence in the upper atmosphere contributes to global warming.

WATER POLLUTION

IN SPITE of the relatively small amount of fresh water available on Earth, the supply should be adequate for the planet. However, an increasing amount is being contaminated so that it is no longer capable of sustaining the life that depends on it. Pollutants in water range from bacteria and viruses in sewage to heavy metals and radioactive substances.

Some water pollution comes from the runoff of rain over farmland, carrying animal wastes, chemical fertilizers and pesticides; and over urban areas, carrying polluting agents from sources ranging from septic tanks to parking lots. This runoff joins surface water in streams and rivers; some also seeps into the ground and joins the deeper water in the water table below the surface. Pollution also comes from factories, power stations and sewage treatment plants, which often dump their various wastes through pipes directly into a body of water such as a river or lake.

Flowing water is more able to tolerate pollution, because its waters are continuously changing over a short period – days or weeks. Still water is less able to recover quickly. The ability of the water to recover also depends on the nature of the polluting agent. Sewage and rotting food are biodegradable – they break down naturally, and relatively quickly. Most industrial wastes, on the other hand, contain chemicals that break down very slowly or not at all. The Rhine in Germany, which flows through some of the most heavily industrialized regions in the world, is one of the most polluted rivers in the world.

Pollution is less visible in the oceans than in smaller bodies of water, simply because they are so vast. The oceans have an enormous capacity for diluting and dispersing wastes, but this is becoming overloaded. Seas and oceans have become the main dumping ground of the world. The Mediterranean receives more than 500 million tonnes a year of sewage from surrounding countries. Two-thirds of this is discharged raw, without even minimal treatment, and contaminates beaches from France to Israel. Swimmers may become ill with a variety of infections; fish, particularly shellfish, die outright or are no longer fit to eat. Plastic wastes and heavy metals such as mercury and lead are also dumped by the thousands of tonnes a year. Massive growths of algae, which flourish

KEYWORDS

ACID RAIN

ALGAL BLOOM

BIODEGRADABLE

BIOLOGICAL OXYGEN
    DEMAND

EFFLUENT

EUTROPHICATION

INDICATOR SPECIES

POLLUTION

RUNOFF

SEWAGE

WATER CYCLE

WATER TABLE

▷ The Rio Tinto in Spain is so called because of the color of its waters (*tinto* means "dark red" or "dyed"). The color comes from what were once Europe's largest copper mines, close to the river. The mining produced a lot of spoil, which piled around the mine. Water running off the site picked up the copper, which colored it. Although the mine itself is now closed, much of the spoil at the site remains, and the huge copper deposits in the river persist. Like other metals, copper is present in many organisms in tiny amounts, but a large dose is extremely toxic.

▽ Water pollution is often difficult to spot. This attractive scene is beside the River Elbe in eastern Germany. However, looks are deceptive: this is one of the most polluted rivers in Europe, carrying pollution from some of Europe's largest industrial centers to the North Sea.

in water polluted by sewage and fertilizers, are washed ashore, clogging coastlines and killing fish. Algal bloom has also struck the coasts of Norway, Sweden and Denmark; it occurs about 50 times a year off the coasts of Japan. In the United States, beaches on the northeast coast were closed in 1988 as medical waste including used hypodermic needles washed ashore. Dumping radioactive waste at sea has also been common practice. It was outlawed in 1983, but non-radioactive industrial waste continues to be deposited.

Every year more than 3 million tonnes of oil pollute the sea. Half of it is from land-based sources such as refineries; a third of it is deliberately dumped when tankers are rinsed. Less than a sixth comes from accidents involving oil tankers. Oil does not dissolve in water, but floats on or near the surface. Birds' feathers become coated and lose their waterproofing qualities, as does the fur of marine animals such as seals. If a bird is covered in oil, it is unable to fly and drowns; if it ingests any oil, its gut can be damaged. Detergents used as oil dispersants are just as damaging.

Acid rain has "killed" thousands of lakes in northern Europe, particularly Sweden. It forms from air pollutants such as sulfur dioxide. When the rain falls on lakes, most forms of life quickly die, although the water is crystal clear.

▽ **As a river flows from its source to the ocean, there are many opportunities for its waters to become polluted. Farmland along the river valley may be treated with chemical fertilizers which, if not applied carefully, end up in the water. Cities are often located along the banks of a river and deal with wastes such as sewage by simply piping them into the water. Industries are also built along the banks, attracted by the supply of water for industrial processes and the ease of transporting goods. They often discharge chemicals that pollute the water. Even plain hot water discharged into the river from power stations lowers the water's oxygen content, killing the fish.**

◁ **Indicator species give ecologists an instant assessment of water pollution. Caddis fly larva, water shrimp s and mayfly nymphs FAR LEFT need clean water. Bloodworms, rat-tailed maggots and tubifex worms LEFT can all survive even if the water is heavily polluted.**

Mayfly nymphs

Rat-tailed maggots

Water shrimp

Tubifex worms

Caddisfly larvae

Bloodworm

Oil spill

Industrial waste

Applying fertilizer by air

City waste

DISAPPEARING FORESTS

Faster than any other biomes, the world's forests are disappearing. As much as a third of the total tree cover has been lost since agriculture began some 10,000 years ago. The remaining forests are home to more species than any other biome, making them the Earth's chief resource for the biodiversity of species.

Tropical rain forests once covered 12 percent of the land on the planet. As well as supporting at least 50 percent of the world's species of plants and animals, they are home to millions of people. But there are other demands on the rainforest: much is cut for timber, especially hardwoods such as teak and mahogany. The land is cleared for oil prospecting, roadbuilding, or to provide pastureland for cattle and for growing crops such as coffee, cocoa and bananas. By the 1990s, less than half of the Earth's original rainforests remained, and clearance continued at a rate of up to 20 million hectares a year.

Forests outside the tropics are suffering the same fate. Huge tracts of boreal forest in Siberia are being cleared for fuel and mining operations. In North America there is a replanting program, but the new conifer forests are not as diverse as the ancient temperate rainforest they are replacing. However, this does not mean that these forests are not important to the global ecology. Temperate forests are the Earth's largest land-based source of carbon. If the trees are cut

KEYWORDS

BIODIVERSITY
DEFORESTATION
DESERTIFICATION
GLOBAL WARMING
GREENHOUSE EFFECT
EROSION

down, they cease to use carbon dioxide in the atmosphere (for photosynthesis). As a result, levels of the gas increase, contributing to the global greenhouse effect.

In Europe and North America, pollution is an even bigger threat than clearing. Millions of hectares of trees in 21 countries are dead or dying. The same damage is occurring in China, where 90 percent of Sichuan province's forests have died. The cause appears to be a combination of pollutants, including acid rain. Trees that do survive are more vulnerable to disease, harsh weather and other stress.

△ In India, local women began a forest-protection scheme called the Chipko movement (chipko means "to hug" or "to cling to"). As loggers arrive to clear a forest, villagers simply encircle the trees, refusing to let them be cut.

◁ A red river of silt runs through cleared tropical forest in Brazil. The tree cover in a forest acts as a moisture retainer; when it is removed, rain runs off the ground rather than being absorbed. The area is then more prone to flooding and to soil erosion, reducing its fertility.

△ Ancient coniferous forests along Canada's Pacific coast have been felled to supply the pulp and paper industry, often to make products as trivial as packaging. Most of the trees used for pulp today are grown on plantations, and are about 20 years old; other trees will be allowed to mature before being cut for timber. The logs are floated downriver to the coast where the pulp mills are located.

The loss of both temperate and tropical forests has alarming implications for the world's climate. Trees pump huge amounts of water into the atmosphere as they transpire. The water vapor condenses, forming clouds, which reflect the Sun's heat – heat that would otherwise reach the Earth's surface and increase its temperature, a phenomenon called global warming. When huge tracts of forests are cleared, rainfall in those areas diminishes, causing the climate to become hotter and drier. The soil also becomes dry and hard, making it less able to absorb rain; its fertility decreases along with its moisture. Eventually, if no restoration is attempted, the forest is transformed into desert.

▷ Forests once covered about 6 billion hectares of the Earth; about 4 billion hectares remain – three times more than pasture or farmland. The largest total area of forest is found in northern Eurasia. South America has the largest remaining rainforest, but also the highest reported rate of deforestation – four times higher than Asia, which may lose half its forests by 2000.

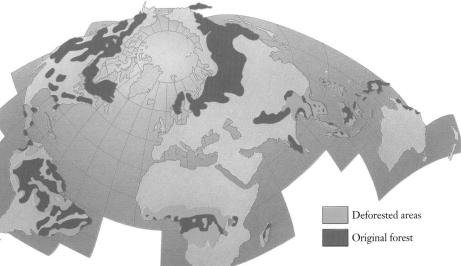

Deforested areas

Original forest

△ Industrial activities such as quarrying and mining leave huge areas of derelict land. In western Australia, a bauxite mine TOP is landscaped to minimize water runoff and soil erosion, and planted with treees and other young plants to restore it. ABOVE Pine trees grow on the site of the former mine. When vegetation cover has been established, other species colonize the ground and a new community develops.

UNDER THREAT OF EXTINCTION

Dinosaurs, saber-toothed tigers and the dodo bird are famous examples of animals that have become extinct. In the case of the dinosaurs, it seems likely that a catastrophic event (probably a meteorite strike) altered the global climate enough to lead to their disappearance. More recent extinctions and near-extinctions – such as the blue whale, tiger, panda and North American bison – have been the direct result of human activity. By the early 1990s, species were becoming extinct at a rate of three per hour, or 27,000 every year – a figure quoted by the American biologist Edward O. Wilson of Harvard University, based on his most conservative estimates. This rate of extinctions carries with it some terrible consequences. Each plant that becomes extinct, for example, may take with it as many as 30 insects and animals that depend on it for food.

Habitat loss is one of the most important causes of extinction. As rising populations in many countries lead to the clearing of more land, habitats such as rainforests and grasslands disappear. In East Africa, once renowned for its wildlife, few wild animals survive outside the boundaries of national parks and game reserves. In other parts of the world, coastal ecosystems are cleared for development. Wetland areas such as the Cota Doñana in southern Spain and the Everglades in Florida are drying out as a result of water extraction to support farming and tourism. Bird species are among the worst affected by the loss of wetlands.

KEYWORDS

ENDANGERED SPECIES
EVOLUTIONARY BIOLOGY
EXTINCTION
GENE POOL
HABITAT LOSS
INBREEDING
OVEREXPLOITATION
POPULATION DYNAMICS
SPECIES

Some species are more vulnerable to extinction than others. A specialized habitat or diet restricts a species to certain locations or foods, especially if it feeds at the upper levels of a food chain, as eagles do. Species with low rates of reproduction, such as blue whales, whooping cranes and giant pandas, may not breed fast enough to keep up with any increase in their death rate, or to keep their populations large enough to avoid inbreeding, which weakens the gene pool and eventually leads to extinction. Large species, such as African lions, elephants and grizzly bears, are often

△ The black rhino was once widespread in central and southern Africa. Since the 1970s its population has plunged from as many as 30,000 to fewer than 3,000 – a result of poaching for its horn. Rhinoceros horn is used in traditional medicine throughout much of Asia and in ceremonial daggers in Yemen.

▽ Exotic insects such as some butterflies are much sought after by collectors. There are well over 2 million insect species, but many have become extinct or are threatened due to this form of hunting.

THE LAST TASMANIAN WOLF

The largest marsupial carnivores, Tasmanian wolves lived in Australia and New Guinea in prehistoric times, but by the 19th century they were found only in Tasmania; competition with the dingo may have eliminated them in Australia. Bounty hunting of Tasmanian wolves began about 1830, because the animals were preying on flocks of sheep introduced by European settlers. By 1850 its survival was threatened, but bounty hunting continued until 1909. The last Tasmanian wolf died in captivity in 1936.

The Siberian tiger RIGHT and the mountain gorilla of central Africa BELOW are among the most endangered species on Earth, as a result of forest clearance (and hunting, in the case of the tiger). There are thought to be fewer than 500 of each left – below the minimum figure for adequate reproduction.

the target of commercial and sport hunters. Animals that prey on livestock or have a reputation for attacking humans, such as wolves and some crocodiles, may be shot on sight. Still others may behave in unusual ways that make them vulnerable. Redheaded woodpeckers fly in front of cars. Key deer are hit by cars as they forage on highways for cigarette butts.

Alien species such as rabbits, rats, dogs and cats, brought by humans to new habitats, have led to the disappearance of many native species. In New Zealand, flightless birds were killed by rats that escaped from ships carrying European colonists. Tourists in pursuit of the wonders of nature also endanger the balance of natural systems: on beaches in Malaya, hundreds of people wait for the endangered giant leather-backed turtle to lay eggs. Their noise and flash cameras frighten the turtles and interfere with the egg-laying. Remote areas such as the Antarctic and Mount Everest are now regularly visited by tourists who trample vegetation and disturb bird colonies.

△ Tourists walk over a section of the Great Barrier Reef off the northeast coast of Australia. The 2000-km-long reef is composed of 2500 individual reefs made up of some 400 species of coral, and supports more than 5500 other marine species. Although large tracts of the reef are restricted, it is difficult to control access in shallow water, where the most damage may be done.

CONSERVING AND RESTORING

I N 1987 the Sierra Club of the United States sponsored a survey of the world's remaining wilderness – defined as any completely undeveloped area of at least 405,000 hectares. About 34 percent of the Earth's land meets this definition. Most of it is forest, desert and tundra, and most is found at high latitudes. However, less than 4 percent of the Earth is protected. With ever more of the world's biomes damaged or disappearing, it is essential that what remains is conserved.

KEYWORDS

BIODIVERSITY
CONSERVATION
ECOLOGY
HABITAT
RESTORATION
SUSTAINABLE

Conservation is inbuilt in the lifestyles and beliefs of most indigenous peoples. In advanced civilizations, conservation began when rulers of countries set aside land for hunting or other royal recreation. The first national park in the world was Yellowstone in the northwestern United States, established in 1872. As the American frontier "closed", it became apparent that the vast country's resources were not infinite, as they had seemed to the first settlers 200 years earlier. Restricted hunting seasons had been introduced in the original 13 colonies as early as 1700 to protect deer.

Now there are more than 1000 national parks in 120 countries around the world, almost half of them

established since 1972; new acreage is continuously being added. Many of these parks were originally introduced to preserve landscapes, but it is now recognized as more important to preserve whole ecosystems because each is important on the global scale. This is far from being achieved. More than half of tropical countries, whose biomes are under greatest threat, have no systematic approach to conservation. (Costa Rica and Botswana are notable exceptions). The conservation of one biome or group of species may be emphasized over general conservation; in southern Africa, most protected areas are those that contain large numbers of mammals, while other areas – equally valuable – are ignored.

Conservation is often hampered by social and economic conditions. Poor countries' priorities are often development and education – or simply food, housing and public health – at the expense of the environment.

△ 1.8 million hectares of Amazonian rainforest have been handed over by Colombia's government to its indigenous population in recognition of their conservation skills. The local people live in and farm the forest. Their traditional shifting cultivation – farming a small patch at a time – does not over-use local resources. If taken up elsewhere, this approach to conservation could save the world's rainforests.

△ The Yosemite valley in California was set aside in 1890 as a national park, one of the first in the world. National parks were created primarily to protect the landscape rather than the wildlife. Ironically, one of the greatest threats to Yosemite is posed by the 3 million tourists who flock there every year.

▷ To stop the killing of elephants, 100 countries have banned trade in ivory. Confiscated tusks made a bonfire in Nairobi National Park, Kenya. The trade has collapsed, and elephant populations are recovering.

△ Antarctica is the last place on Earth that is all wilderness. 25 countries have research stations there, but exploration for commercial purposes (such as mining) was banned by international agreement in 1959. In 1991 the ban was renewed until 2041. It has been proposed that the entire continent (only 2 percent of which is not under ice) should be made a World Park and all access carefully controlled.

Their economic interests are sometimes combined with conservation programs supported by international organizations. In 1987 the first "debt for nature swap" was arranged. $650,000 worth of Bolivia's national debt was written off against a 160,000-sq-km area of rainforest and savanna. This area is now designated a biosphere reserve: a whole ecosystem with a strictly controlled central zone where no interference is allowed, surrounded by a transition zone in which research is permitted, and finally a buffer zone to protect the ecosystem from encroachment. There are now more than 250 worldwide.

Conservation sites require management in order to be successful. Careful account must be taken of what each species requires for living space and food, and how many individuals make up a viable population. The needs of any local people must also be considered. Villagers in southern Africa are being educated to realize that animal populations are a valuable tourist attraction, and are learning to look after the animals that represent their future. In these areas, poaching of animals such as rhinoceros and elephants has declined sharply, allowing controlled big-game hunting by paying tourists to be part of conservation strategy.

Restoring derelict areas is another increasingly important aspect of conservation. Quarry faces may become inland cliffs covered in climbing plants. Gravel pits can be flooded and carefully shaped to attract waterfowl. It is even possible to grow plants on the huge spoil tips associated with heavy mining. Plants are being bred to tolerate the high levels of heavy metals that are present in the spoil. One of the greatest challenges is to restore the scarred landscape created by open-cast mining.

141

PRESERVING DIVERSITY

As more species become extinct each year, awareness has grown of the importance of preserving wildlife. Ecologists have given a new understanding of the contribution made by most species to the ecosystems in which they live. Other scientists have discovered many benefits to humans to be derived from threatened species.

Only a fraction of the world's plants and animals have been studied for their potential value as food, medicine or materials for industry. Many species may become extinct without revealing their potential value. Each one is a storehouse of natural genetic and biochemical resources that cannot be replaced.

As commercial agriculture spreads throughout the world, much of the traditional variety of food crops has been lost. Three types of grain – wheat, rice and corn (maize) – make up half the world's food harvest, and many modern crops have lost their genetic diversity. A crop with nearly uniform genetic composition may become susceptible to a particular disease, which can wipe out the entire crop. The most effective way of preventing disease among such crops is to interbreed them with wild strains.

Interbreeding crops with wild plants also increases yields and expands the area suitable for cultivation. For example, 10 million square kilometers on Earth is too salty for growing the species of grain now used in

▷ Zoos and aquaria have been criticized for keeping wild animals in cramped, unsuitable conditions, for concentrating on a few spectacular species and for being more concerned with entertainment than with education or real conservation. Many zoos have attempted to address these issues, and their role in preserving the most endangered species has grown in recent years.

△ Lemurs are found only on the tropical island of Madagascar off the east coast of Africa. Rainforest clearance and hunting have taken their toll and only 30 of the original 45 species survive. One of the most threatened is the aye-aye, a shy nocturnal lemur. The locals considered it an evil animal and they killed it on sight. By the mid 1960s it was on the brink of extinction. In 1966, in an attempt to save the species from extinction, nine aye-aye were moved to an island off the coast of Madagascar. The aye-aye survives with a very small population.

◁ Przewalski's horse is the last surviving species of wild horse, which became extinct in its native habitat of the central Asian grasslands. Small herds have been maintained in several Western zoos, and a record of the pedigree of each individual has been maintained. This enabled scientists to minimize inbreeding, and to keep the gene pool as wide as possible. In the 1990s attempts were made to reintroduce the species to its natural habitat.

intensive farming, but would be suited to the strains of wild wheat, rice, barley and millet that prefer salty soil.

National gene banks now exist in more than 60 countries in order to attempt to preserve the genetic diversity of important food crops. Sixty thousand varieties of rice and twelve thousand types of wheat and corn from 47 countries are stored as seed samples. However, seeds cannot be stored infinitely; they deteriorate and are vulnerable to disease.

Tropical rain forests have the world's greatest biodiversity, making them a storehouse of opportunity for biotechnology, medicine, agriculture and horticulture. More than ten percent of all common medicines originate in tropical rainforests, and the potential for future discoveries is evident from the fact that 6500 plants from the rain forests of southeast Asia alone are used for medicinal purposes.

As the numbers of large mammals decline, zoos take on the role of animal gene banks. When endangered species are bred in captivity, animals with the widest possible genetic makeup are used in order to keep the gene pool as large as possible. Over the next 50–100 years, many of the larger mammals may cease to exist in the wild. The only populations will live in zoos where they can be bred and maintained, to be appreciated and valued by future generations.

◁ There are 250,000 species of flowering plants in the world. Of these at least 20,000 have edible parts, but only 3000 have been used by people. The great diversity among the flowering plants must be preserved. The Royal Botanic Gardens in Kew, London, has one of the most important plant collections in the world.

These date back more than 200 years, and comprise more than five million dried plant specimens, leaves, seeds, seed pods and fruits, covering 5000 species of wild plant. Some plants, such as the cacao, can be kept only by growing specimens. Similar plant libraries are kept for key commercial plants, such as rice, wheat and potatoes.

△ Diversity is best maintained in the wild by preserving habitats. On temperate agricultural land, hedgerows provide a haven for wildlife, which once had lived in the ancestral broadleaved forest. As agriculture becomes more mechanized, such hedges are lost and many species of insects, plants and mammals are endangered.

FACTFILE

PRECISE MEASUREMENT is at the heart of all science, and the several standard systems have been in use in the present century in different societies. Today, the SI system of units is universally used by scientists, but other units are used in some parts of the world. The metric system, which was developed in France in the late 18th century, is in everyday use in many countries, as well as being used by scientists; but imperial units (based on the traditional British measurement standard, also known as the foot–pound–second system), and standard units (based on commonly used American standards) are still in common use.

Whereas the basic units of length, mass and time were originally defined arbitrarily, scientists have sought to establish definitions of these which can be related to measurable physical constants; thus length is now defined in terms of the speed of light, and time in terms of the vibrations of a crystal of a particular atom. Mass, however, still eludes such definition, and is based on a piece of platinum-iridium metal kept in Sèvres, near Paris.

☐ METRIC PREFIXES

Very large and very small units are often written using powers of ten; in addition the following prefixes are also used with SI units. Examples include: milligram (mg), meaning one thousandth of a gram, kilogram (kg), meaning one thousand grams.

Name	Number	Factor	Prefix	Symbol
trillionth	0.000000000001	10^{-12}	pico-	p
billionth	0.000000001	10^{-9}	nano-	n
millionth	0.000001	10^{-6}	micro-	μ
thousandth	0.001	10^{-3}	milli-	m
hundredth	0.01	10^{-2}	centi-	c
tenth	0.1	10^{-1}	deci-	d
one	1.0	10^{0}	–	–
ten	10	10^{1}	deca-	da
hundred	100	10^{2}	hecto-	h
thousand	1000	10^{3}	kilo-	k
million	1,000,000	10^{6}	mega-	M
billion	1,000,000,000	10^{9}	giga-	G
trillion	1,000,000,000,000	10^{12}	tera-	T
quadrillion	1,000,000,000,000,000	10^{15}	exa-	E

☐ CONVERSION FACTORS

Conversion of METRIC units to imperial (or standard) units

To convert:	to:	multiply by:
LENGTH		
millimeters	inches	0.03937
centimeters	inches	0.3937
meters	inches	39.37
meters	feet	3.2808
meters	yards	1.0936
kilometers	miles	0.6214
AREA		
square centimeters	square inches	0.1552
square meters	square feet	10.7636
square meters	square yards	1.196
square kilometers	square miles	0.3861
square kilometers	acres	247.1
hectares	acres	2.471
VOLUME		
cubic centimeters	cubic inches	0.061
cubic meters	cubic feet	35.315
cubic meters	cubic yards	1.308
cubic kilometers	cubic miles	0.2399
CAPACITY		
milliliters	fluid ounces	0.0351
milliliters	pints	0.00176 (0.002114 for US pints)
liters	pints	1.760 (2.114 for US pints)
liters	gallons	0.2193 (0.2643 for US gallons)
WEIGHT		
grams	ounces	0.0352
grams	pounds	0.0022
kilograms	pounds	2.2046
tonnes	tons	0.9842 (1.1023 for US, or short, tons)
TEMPERATURE		
Celsius	fahrenheit	1.8, then add 32

Conversion of STANDARD (or imperial) units to metric units

To convert:	to:	multiply by:
LENGTH		
inches	millimeters	25.4
inches	centimeters	2.54
inches	meters	0.245
feet	meters	0.3048
yards	meters	0.9144
miles	kilometers	1.6094
AREA		
square inches	square centimeters	6.4516
square feet	square meters	0.0929
square yards	square meters	0.8316
square miles	square kilometers	2.5898
acres	hectares	0.4047
acres	square kilometers	0.00405
VOLUME		
cubic inches	cubic centimeters	16.3871
cubic feet	cubic meters	0.0283
cubic yards	cubic meters	0.7646
cubic miles	cubic kilometers	4.1678
CAPACITY		
fluid ounces	milliliters	28.5
pints	milliliters	568.0 (473.32 for US pints)
pints	liters	0.568 (0.4733 for US pints)
gallons	liters	4.55 (3.785 for US gallons)
WEIGHT		
ounces	grams	28.3495
pounds	grams	453.592
pounds	kilograms	0.4536
tons	tonnes	1.0161
TEMPERATURE		
fahrenheit	Celsius	subtract 32, then × 0.55556

□ SI UNITS

Now universally employed throughout the world of science and the legal standard in many countries, SI units (short for *Système International d'Unités*) were adopted by the General Conference on Weights and Measures in 1960. There are seven base units and two supplementary ones, which replaced those of the MKS (meter–kilogram–second) and CGS (centimeter–gram–second) systems that were used previously. There are also 18 derived units, and all SI units have an internationally agreed symbol.

None of the unit terms, even if named for a notable scientist, begins with a capital letter: thus, for example, the units of temperature and force are the kelvin and the newton (the abbreviations of some units are capitalized, however). Apart from the kilogram, which is an arbitrary standard based on a carefully preserved piece of metal, all the basic units are now defined in a manner that permits them to be measured conveniently in a laboratory.

Name	Symbol	Quantity	Standard
BASIC UNITS			
meter	m	length	The distance light travels in a vacuum in $\frac{1}{299,792,458}$ of a second
kilogram	kg	mass	The mass of the international prototype kilogram, a cylinder of platinum-iridium alloy, kept at Sèvres, France
second	s	time	The time taken for 9,192,631,770 resonance vibrations of an atom of cesium-133
kelvin	K	temperature	$\frac{1}{273.16}$ of the thermodynamic temperature of the triple point of water
ampere	A	electric current	The current that produces a force of 2×10^{-7} newtons per meter between two parallel conductors of infinite length and negligible cross section, placed one meter apart in a vacuum
mole	mol	amount of substance	The amount of a substance that contains as many atoms, molecules, ions or subatomic particles as 12 grams of carbon-12 has atoms
candela	cd	luminous intensity	The luminous intensity of a source that emits monochromatic light of a frequency 540×10^{-12} hertz and whose radiant intensity is $\frac{1}{683}$ watt per steradian in a given direction
SUPPLEMENTARY UNITS			
radian	rad	plane angle	The angle subtended at the center of a circle by an arc whose length is the radius of the circle
steradian	sr	solid angle	The solid angle subtended at the center of a sphere by a part of the surface whose area is equal to the square of the radius of the sphere

Name	Symbol	Quantity	Standard
DERIVED UNITS			
becquerel	Bq	radioactivity	The activity of a quantity of a radio-isotope in which 1 nucleus decays (on average) every second
coulomb	C	electric current	The quantity of electricity carried by a charge of 1 ampere flowing for 1 second
farad	F	electric capacitance	The capacitance that holds charge of 1 coulomb when it is charged by a potential difference of 1 volt
gray	Gy	absorbed dose	The dosage of ionizing radiation equal to 1 joule of energy per kilogram
henry	H	inductance	The mutual inductance in a closed circuit in which an electromotive force of 1 volt is produced by a current that varies at 1 ampere per second
hertz	Hz	frequency	The frequency of 1 cycle per second
joule	J	energy	The work done when a force of 1 newton moves its point of application 1 meter in its direction of application
lumen	lm	luminous flux	The amount of light emitted per unit solid angle by a source of 1 candela intensity
lux	lx	illuminance	The amount of light that illuminates 1 square meter with a flux of 1 lumen
newton	N	force	The force that gives a mass of 1 kilogram an acceleration of 1 meter per second per second
ohm	Ω	electric resistance	The resistance of a conductor across which a potential of 1 volt produces a current of 1 ampere
pascal	Pa	pressure	The pressure exerted when a force of 1 newton acts on an area of 1 square meter
siemens	S	electric conductance	The conductance of a material or circuit component that has a resistance of 1 ohm
sievert	Sv	dose	The radiation dosage equal to 1 joule equivalent of radiant energy per kilogram
tesla	T	magnetic flux density	The flux density (or density induction) of 1 weber of magnetic flux per square meter
volt	V	electric potential	The potential difference across a conductor in which a constant current of 1 ampere dissipates 1 watt of power
watt	W	power	The amount of power equal to a rate of energy transfer of (or rate of doing work at) 1 joule per second
weber	Wb	magnetic flux	The amount of magnetic flux that, decaying to zero in 1 second, induces an electromotive force of 1 volt in a circuit of one turn

Many factors combine to determine the climate of a region. These include the latitude (the low-latitude tropics are much warmer than the high-latitude polar regions), prevailing winds, topography (temperature decreases with height) and distance from the sea. Also significant are ocean currents, which in turn result from a combination of the effects of wind, differences in water density caused by changes in temperature and salinity, and the Earth's rotation. The rotation causes the Coriolis effect, which makes winds in the Northern Hemisphere deflect to the right, and those in the Southern Hemisphere deflect to the left.

Even within the four major climate types designated tropical, subtropical, temperate and cold, there are variations resulting from the differences in humidity. For example, in adjacent latitudes, there is a great contrast between the very humid tropical forest climates and the dry climates characteristic of the nearby hot, arid deserts.

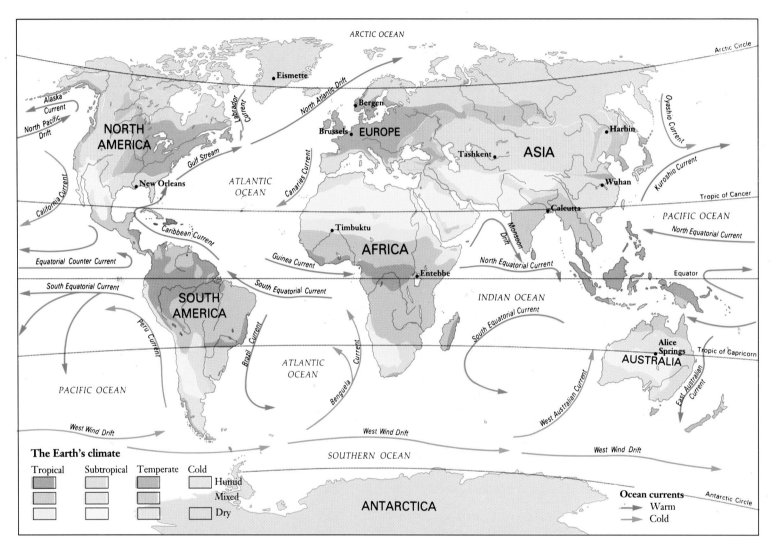

The Earth's climate

Tropical Subtropical Temperate Cold

Humid

Mixed

Dry

Ocean currents
→ Warm
→ Cold

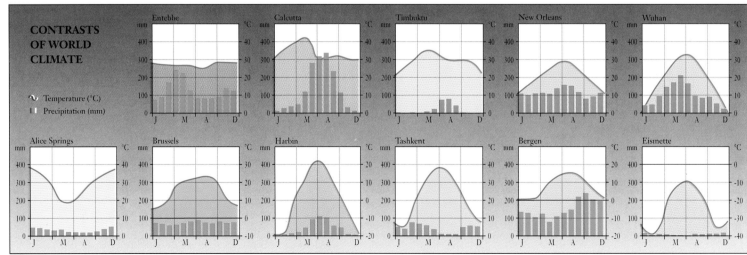

CONTRASTS OF WORLD CLIMATE

∿ Temperature (°C)
▌ Precipitation (mm)

Entebbe · Calcutta · Timbuktu · New Orleans · Wuhan

Alice Springs · Brussels · Harbin · Tashkent · Bergen · Eismette

WORLD TEMPERATURES IN JANUARY

▷ The warmest parts of the world in January are in the southern hemisphere – during the southern summer – while at the same time of the year the northern hemisphere endures its winter. In Siberia, winter temperatures below -65°C have been recorded, while northern Australians swelter at temperatures in excess of 30°C.

Average January temperature

°C
30
20
10
0
-10
-20
-30
-40

Average July temperature

°C
30
20
10
0
-10

WORLD TEMPERATURES IN JULY

◁ This time of the year corresponds to the northern summer, although in the southern hemisphere temperatures outside Antarctica seldom anywhere fall below -10°C. The hottest regions are northern and Saharan Africa, the Middle East and northern India. The hottest temperature on record, 58°C, was recorded in Libya.

WORLD PRECIPITATION

▷ The wettest places on Earth are in the tropics where, in some rain forests, annual rainfall in excess of 3000 millimeters is common. Paradoxically, many of the driest regions are in the next band of latitudes and in the centers of continents, remote from the sea. In polar regions, most of the precipitation falls as snow. The graphs LEFT compare temperature and precipitation in various locations.

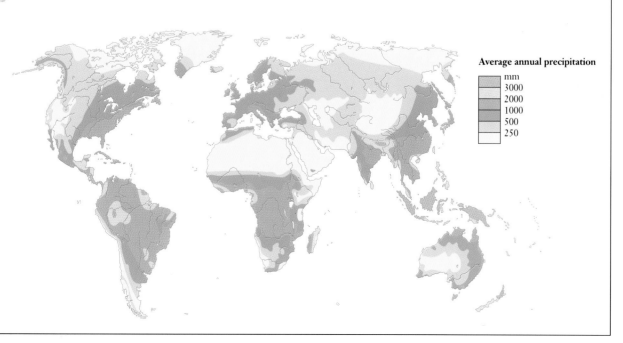

Average annual precipitation

mm
3000
2000
1000
500
250

The world's farmland can be divided into five zones: arable (cereal crops); fruit, vegetables and tree crops; pasture (mainly for cattle and sheep); rough grazing; and woods and forest (for timber). The large map on these pages identifies these zones, along with nonagricultural land and the world's major fishing grounds. The smaller maps show the distribution of individual farm products.

The chief determining factors for a particular crop or farm animal are the climate – temperature and rainfall – and the nature of the soil. For example, arable crops need well-drained fertile soils and plenty of rain, and this type of farming is concentrated in temperate regions (for wheat) and wet subtropics (for rice). Livestock farming, on the other hand, can be carried out in drier conditions; goats are particularly tolerant of arid terrain.

Rice and wheat

Rice
Wheat
Mixed

Millet and sorghum, oats and rye

Millet and sorghum
Oats and rye
Mixed

Cattle

Cattle

Sheep and goats

Sheep and goats

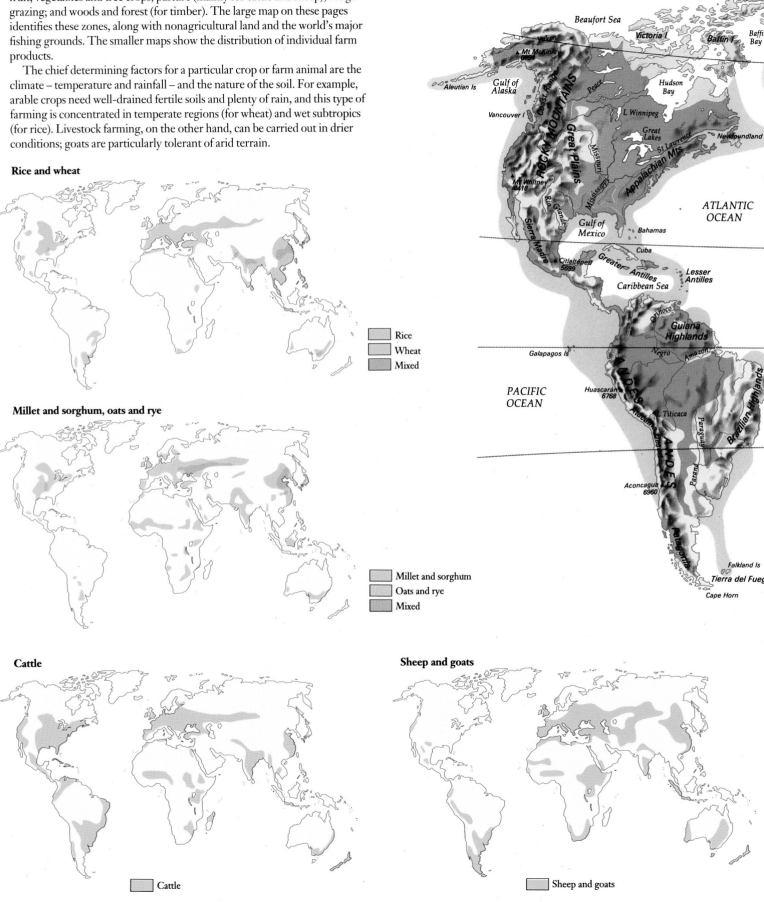

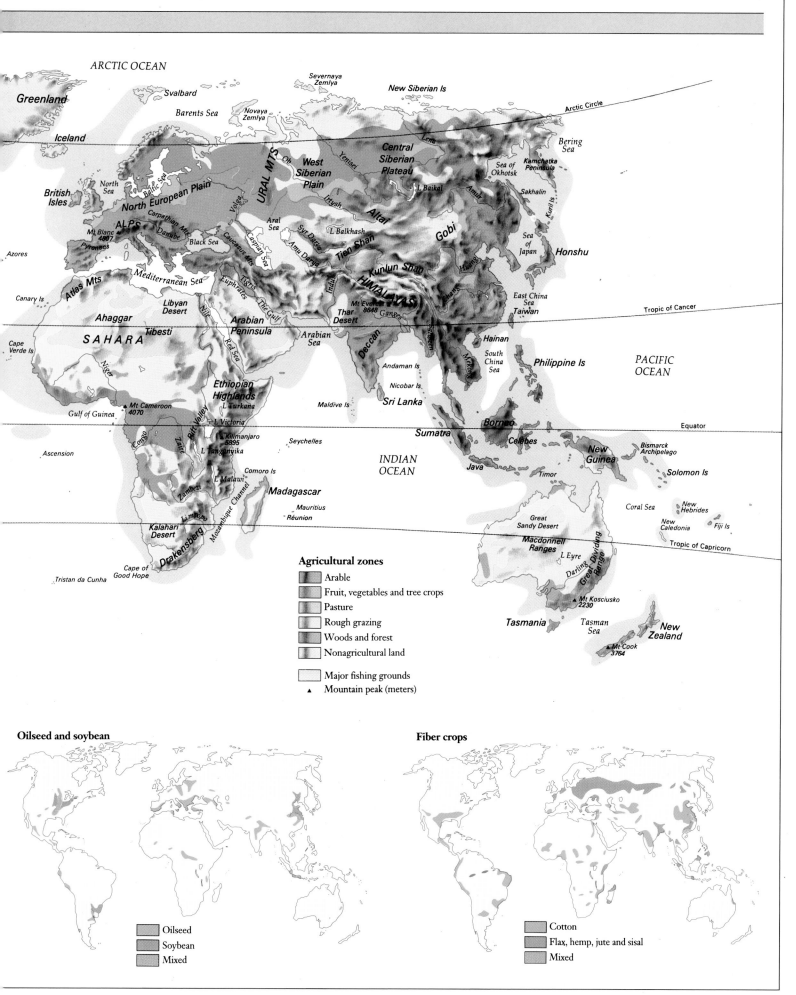

ARCTIC OCEAN

Greenland

Svalbard

Severnaya
Zemlya

New Siberian Is

Iceland

Barents Sea

Novaya
Zemlya

Arctic Circle

British
Isles

North
Sea

Baltic Sea

North European Plain

URAL MTS

Ob

West
Siberian
Plain

Yenisei

Central
Siberian
Plateau

Lena

Bering
Sea

Kamchatka
Peninsula

Sea of
Okhotsk

Sakhalin

ALPS

Carpathian Mts

Volga

Aral
Sea

Irtysh

L Baikal

Amur

Kuril Is

Mt Blanc
4807

Danube

Pyrenees

Black Sea

Caucasus Mts

Caspian Sea

Syr Darya

L Balkhash

Altai

Tien Shan

Gobi

Sea
of
Japan

Honshu

Azores

Amu Darya

Kunlun Shan

Sea of Japan

Canary Is

Atlas Mts

Mediterranean Sea

Euphrates

Tigris

The Gulf

Indus

HIMALAYAS

Mt Everest
8848

Ganges

Huang

Yangtze

Salween

East China
Sea

Taiwan

Tropic of Cancer

Cape
Verde Is

SAHARA

Ahaggar

Tibesti

Libyan
Desert

Nile

Red Sea

Arabian
Peninsula

Arabian
Sea

Thar
Desert

Deccan

Hainan

South
China
Sea

Philippine Is

PACIFIC
OCEAN

Niger

Ethiopian
Highlands

L Turkana

Andaman Is

Nicobar Is

Sri Lanka

Mekong

Gulf of Guinea

Mt Cameroon
4070

Rift Valley

Zaire

L Victoria

Kilimanjaro
5895

Maldive Is

Borneo

Equator

Ascension

Congo

L Tanganyika

Seychelles

Sumatra

Celebes

New
Guinea

Bismarck
Archipelago

INDIAN
OCEAN

Java

Timor

Solomon Is

L Malawi

Comoro Is

Mauritius

Coral Sea

New
Hebrides

Zambezi

Madagascar

Réunion

Great
Sandy Desert

New
Caledonia

Fiji Is

Limpopo

Mozambique Channel

Kalahari
Desert

Drakensberg

Macdonnell
Ranges

L Eyre

Great Dividing Range

Darling

Tropic of Capricorn

Cape of
Good Hope

Agricultural zones

Tristan da Cunha

Mt Kosciusko
2230

Tasmania

Tasman
Sea

New
Zealand

Arable

Fruit, vegetables and tree crops

Pasture

Rough grazing

Mt Cook
3764

Woods and forest

Nonagricultural land

Major fishing grounds

▲ Mountain peak (meters)

Oilseed and soybean

Fiber crops

Oilseed

Soybean

Mixed

Cotton

Flax, hemp, jute and sisal

Mixed

People have been exploiting the world's mineral resources since the rise of humankind. Indeed, the names of those remote eras reflect the materials used: the Stone Age, the Bronze Age and then the Iron Age. Even before that time, early humans used wood – and later charcoal – as a fuel. Much later, fossil fuels such as coal were exploited, followed in the last century or so by oil (petroleum) and natural gas.

Wood is a renewable source because more trees can be planted to replace those that are felled – as long as they are replaced at least as fast as they are cut down. But fossil fuels are not renewable: once burned they are gone for ever. At the present time, fossil fuels provide all by 7 percent of the world's energy needs: oil accounts for 45 percent, coal for 29 percent and natural gas for 19 percent (the remainder is provided by hydroelectric sources and by nuclear power). But this picture will change as reserves of fuels dwindle. Estimates vary widely, but it is though that there are currently 700 billion barrels of oil in reserves worldwide – 43 percent of it in the Middle East. Coal reserves in the late 1980s were estimated at 1.2 trillion tonnes – 43 percent of it in China.

Mineral resources, too, are being used up. For example, there is currently enough iron ore left to last for only about 165 years at the present rate of consumption, and only enough copper to last for 40 years. Like fossil fuels, these resources cannot be replaced once they are used. But scrap metals can be collected and recycled – 30 percent of each year's aluminum production comes from recycled scrap, and 50 percent of steel production. Glass and plastics can also be recycled, saving both their mineral content and much of the energy that goes into manufacturing them.

◻ **MINERAL RESOURCES**

Mineral	Uses	Occurrence
Bauxite	Aluminum is used in foil, packaging and cans, sports gear, chair frames, ladders	As aluminum oxide, the chief ore of aluminum, bauxite is found in USA,Canada, Russia, Brazil and Australia
Copper	Electrical wiring, alloys such as brass and bronze, paint pigments, insecticides	Mostly as copper pyrites (copper sulfide) in Chile, USA, Canada, Russia, Zambia and Zaire
Diamonds	Jewelry; non-gem quality South Africa, Brazil and India and for cutting tools	Gem-quality diamonds mined in stones are are used as abrasives
Gold	Jewelry, coins/bullion, plating electrical contacts	Occurs as the free metal underground and as particles in alluvial deposits, mainly in South Africa, Russia, Canada, USA and Australia
Iron ore	Main use of iron remains manufacture of steel, with some for cast and wrought iron	Chief ores are hematite and magnetite (iron oxides), found mainly in Russia,Brazil, China, Australia, Canada, South Africa, Sweden and France
Potash	Fertilizers, manufacture of explosives	Potassium-containing salts include potassium carbonate and chloride, found chiefly in China, Australia, USA, East Africa and Central Europe
Tin	Coating steel for food and drinks cans, and for making alloys such as solder and pewter	Chief ore is cassiterite (tin oxide) fromMalaysia, Peru, Russia, USA, Canada, Poland and Australia

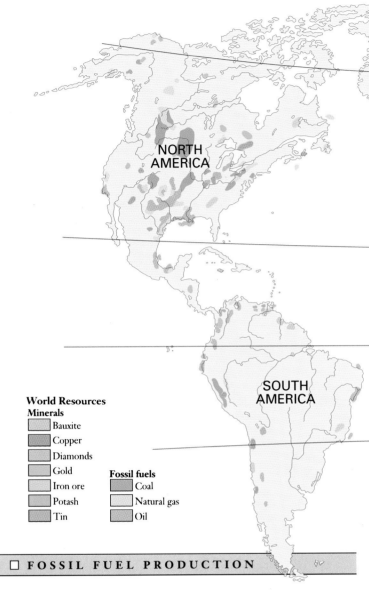

World Resources
Minerals
Bauxite
Copper
Diamonds
Gold
Iron ore
Potash
Tin

Fossil fuels
Coal
Natural gas
Oil

◻ **FOSSIL FUEL PRODUCTION**

MAIN NATURAL GAS PRODUCERS

Country	m³ per year
Russia	649 million
United States	487
Canada	96
Netherlands	81
Great Britain	46
Algeria	44
Romania	36
Indonesia	35
Mexico	34

MAIN COAL PRODUCERS

Country	Tonnes per year
China	956 million
United States	862
CIS	785
Germany	500
Poland	284

MAIN OIL PRODUCERS

Country	Barrels per year
CIS	4.4 billion
United States	2.8
Saudi Arabia	1.8
Iran	1.0
Iraq	1.0
China	0.96
Mexico	0.85

WORLD OIL PRODUCTION

Year	Barrels per day
1960-64	122.0 million
1965-69	179.1
1970-74	257.3
1975-79	293.9
1980-84	275.2
1985-89	280.7
1990-92*	177.1

* 3-year period; 1 barrel = 159 liters

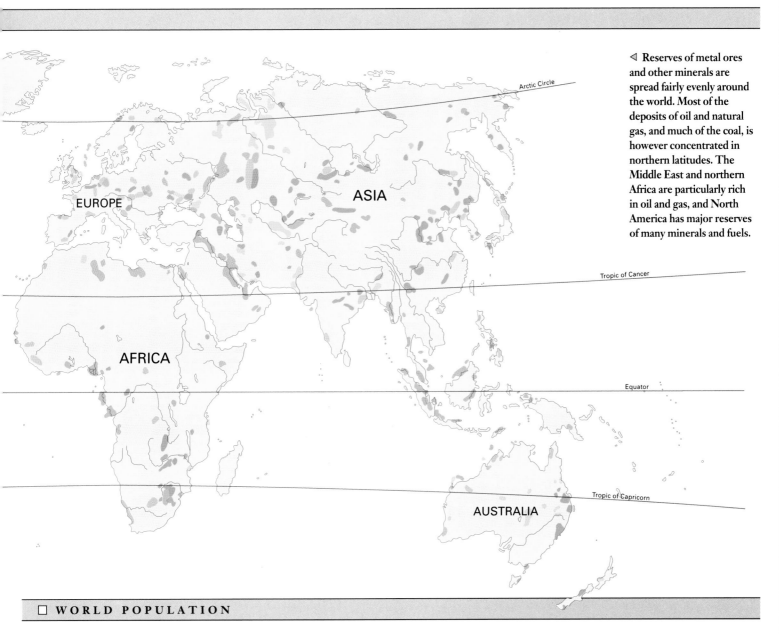

◁ Reserves of metal ores and other minerals are spread fairly evenly around the world. Most of the deposits of oil and natural gas, and much of the coal, is however concentrated in northern latitudes. The Middle East and northern Africa are particularly rich in oil and gas, and North America has major reserves of many minerals and fuels.

□ WORLD POPULATION

People constitute one of the world's major resources, but they bring problems of their own to the environment, mainly because of their rapidly increasing numbers. The world's population more than doubled (from 2.5 billion to 5.3 billion) in the 40 years between 1950 and 1990. By the year 2000 it is expected to reach 6.3 billion, rising to 9 or 10 billion over the next hundred years.

These forecasts of more and more people to be supported by the world's diminishing resources depend not on rising birth rates, but on increased life expectancy. At present, life expectancy is highest in the industrialized nations of the world. But better nutrition and medical treatment will mean that fewer people in developing countries will die young, a fact confirmed by the rapidly declining rates of child mortality in these countries. However, there are still considerable gaps between developed and developing countries in life expectancy and child mortality.

World food production is sufficient for the population; there are even large surpluses in industrialized countries. But many people still starve in in developing countries, because local farming has been disrupted (by war; by flood or drought; by attempts to introduce inappropriate largescale farming; or by neglecting food production for cash crops); or because, in cities, the poor cannot afford to buy food that is often imported.

REGIONAL POPULATION DATA

Region	Area (km²)	% of Earth's land area	Population in 1990	Population density (per km²)	Life expectancy (years)	Estimated population in 2000 (millions)
Asia	43,250	30.8	3.13 billion	72	60.1	3.87 billion
Europe/Russia	11,900	8.5	787 million	66	73.6	947 million
Africa	30,700	2.9	642 million	21	51.8	517 million
North America	23,700	16.9	427 million	18	71.9	514 million
South America	17,450	12.4	297 million	17	65.4	390 million
Australasia	13,250	9.4	26.5 million	2	75.0	29 million

CHILD MORTALITY (under 5 year olds)

Years	Deaths per 1000	Industrial World	Developing countries
1956-60	215	51	255
1961-65	182	39	213
1966-70	161	32	184
1971-75	144	26	164
1976-80	131	24	149
1981-85	118	19	134
1986-90	105	17	119

One of the best ways of preserving wildlife is to set aside areas where it can thrive without interference. This is not a new idea – the Yellowstone National Park in the United States (the world's first) was established as long ago as 1872. Today about 3 percent of available land is devoted to this purpose, although the World National Parks Congress of 1982 recommended that it should be increased to 10 percent.

Various types of protected areas are recognized by organizations such as the World Conservation Union and the United Nations, which has an official list of National Parks and Protected Areas. They include Biosphere Reserves, Marine Reserves, National Maritime Parks, National Parks, National Preserves, Nature Reserves, Wildlife Refuges and World Heritage Sites. These sites are about 90 areas designated to be of outstanding value. They were established by the Convention Concerning the Protection of World Cultural and Natural Heritage, which was adopted in 1972 and came into force in December 1975. Nations that sign the convention nominate

sites, which are then evaluated by the World Heritage Committee. Those that qualify are defined as follows: natural features consisting of physical and biological formations of outstanding universal value from an esthetic or scientific point of view; geological or physiographical formations; areas that are the habitat of threatened species of animals and plants of outstanding universal value from the point of view of science or conservation; and natural areas of outstanding universal value from a scientific, conservation or natural beauty point of view.

By 1980 there were approximately 5000 sites on the United Nations List of National Parks and Protected Areas, mainly national parks but including also most World Heritage Sites. They ranged from the vast Northeast Greenland National Park, with an area of 70 million hectares, and the Great Barrier Reef National Park (20 million hectares) to small oceanic islands of only a few hundred hectares. The table on these pages lists all the world's national parks that have an area of more than 70,000 hectares.

Location	Country	Area (hectares)	Location	Country	Area (hectares)	Location	Country	Area (hectares)
AFRICA			Kruger	South Africa	2,000,000	Bandipur	India	87,500
Banc d'Arguin	Mauritania	1,173,000	Mana Pools	Zimbabwe	219,500	Beydaglari	Turkey	70,500
Benoue	Cameroon	180,000	Massa	Morocco	72,000	Daisetsuzan	Japan	231,000
Bikuar	Angola	790,000	Monovo-Gounda-St Floris	Central African Republic	1,740,000	Fuji-Hakone-Izu	Japan	122,500
Boma	Sudan	2,280,000	Moremi	Botswana	390,000	Golestan	Iran	92,000
Bouche du Baoule	Mali	350,000	Namib-Naukluft	Namibia	4,977,000	Kanha	Bangladesh	94,000
Chobe	Botswana	998,000	Niokolo-Koba	Senegal	913,000	Khao Yai	Thailand	217,000
Dinder	Sudan	890,000	Po	Burkina	155,500	Khunjerab	Jammu & Kashmir	227,000
El Kala	Algeria	76,500	South Luangwa	Zambia	905,000	Kirthar	Pakistan	309,000
Etosha	Namibia	2,227,000	Tasssili N'Ajjer	Algeria	300,000	Komodo	Indonesia	75,000
Fazao-Malfakassa	Togo	192,000	Tsavo	Kenya	2,082,000	Mount Apo	Philippines	73,000
Gile	Mozambique	210,000	Virunga	Zaire	780,000	Mount Kinabalu	Brunei	75,500
Hwange (Wankie)	Zimbabwe	1,465,000	Zakouma	Chad	300,000	Namdapha	India	198,500
Iona	Angola	1,515,000	Zinave	Mozambique	500,000	Royal Bardia	India/Nepal	97,000
Isalo	Madagascar	81,500				Royal Chitwan	Nepal	93,000
Kahuzi-Biega	Zaire	600,000	**ASIA**			Ruhana	Sri Lanka	98,000
Kainji Lake	Nigeria	535,000	Akan	Japan	90,500	Sagarmartha	Nepal	115,000
Kalahari Gemsbok	South Africa	959,000	Alaungdaw Kathapa	Burma	160,500	Sundarbans	India	133,000
Kasungu	Malawi	231,500	Asir	Saudi Arabia	415,000	Taman Negara	Malaysia	434,500
Kisama	Angola	996,000	Aso	Japan	72,500	Tanjung Puting	Indonesia	355,000
Korup	Cameroon	126,000	Bandai-Asahi	Japan	189,500			

□ **FURTHER READING**

Allaby, M. *Into Harmony with the Planet* (Bloomsbury, London 1990)

Angel, H. *The Natural History of Britain and Ireland* (Michael Joseph, 1986)

Attenborough, David *The Living Planet* (Collins and BBC Publications, London, 1984)

Audubon Society Nature Guides: *Deserts* (Alfred Knopf Inc., 1987)

Bradshaw, A.D. and Chadwick, M. J. *The Restoration of Land* (Blackwell Scientific Publications, Oxford, 1980)

Burton, J.A. *The Atlas of Endangered Species* (David and Charles, 1992)

Durrell, L. *State of the Ark* (Doubleday, New York, 1986)

Cadogan, A. and Best, G. *Environment and Ecology* (Nelson Blackie, 1992)

Chernov, Y.I. *The Living Tundra* (Cambridge University Press, Cambridge 1985)

Cloudsley-Thompson, J.L. *Terrestrial Environments* (Croom Helm, London, 1984)

Cox, C.B. and Moore, P.D. *Biogeography: An Ecological and Evolutionary Approach* (Blackwell Scientific Publications, Oxford, 1985)

Cox, B. Moore, P.D. and Whitfield P. *The Atlas of the Living World* (Weidenfield and Nicolson, London, 1989)

Crump, A. *Dictionary of Environment and Development* (Earthscan, 1991)

Diamond, A.W., Schreiber R.L., Attenborough D. and Prestt, I. *Save the Birds* (Cambridge University Press, London and New York, 1987)

Dugan, P. (ed) *Wetlands in Danger* (Mitchell Beazley, 1993)

Ehrlich, P. and Ehrlich, A.H. *Extinction: The causes and consequences of the disappearance of species* (Random House, New York, 1981)

Ehrlich, P.R., Ehrlich, A.H. and Holdren, J.P. *Ecoscience: Population, Resources, Environment* (W.H. Freeman, 1977)

Elton, C.S. *The Pattern of Animal Communities* (Chapman and Hall, London, 1966, 1970)

Fothergill, A. *Life in the Freezer* (BBC Books, London, 1993)

Gardiner, B. and Moore, P. (ed) *The Guinness Encyclopedia of the Living World* (Guinness Publishing, 1992)

Green, N.P.O., Stout, G.W., and Taylor, D.J. *Biological Science* Vol. 1 (Cambridge, 1990)

Grime, J.P. *Plant Strategies and Vegetation Processes* (John Wiley and Sons , 1979)

Huxley, A. *Green Inheritance* (William Collins and Sons, London, 1984)

Kimber, G. and Feldman, M. *Wild Wheat: An Introduction* (The Weizmann Institute of Science, Israel, 1987)

Location	Country	Area (hectares)
Ujung Kulon	Indonesia	78,500
Uromiyeh	Iran	1,295,500
Wilpattu	Sri Lanka	132,000
AUSTRALASIA		
Arthur's Pass	New Zealand	94,500
Fiordland	New Zealand	1,252,000
Flinders Ranges	Australia	80,500
Great Barrier Reef	Australia	20,000,000
Kakadu	Australia	667,000
Mount Aspiring	New Zealand	285,500
Nelson Lakes	New Zealand	96,000
Simpson Desert	Australia	692,500
Tongariro	New Zealand	76,500
Uluru (Ayers Rock)	Australia	132,500
Urewera	New Zealand	207,500
Western Tasmania Wilderness	Australia	76,500
Westland/ Mount Cook	New Zealand	187,500
Wet Tropics of Queensland	Australia	920,000
EUROPE AND RUSSIA		
Borgefjell	Norway	108,500
Brecon Beacons	Great Britain	134,500
Cévennes	France	323,000
Dartmoor	Great Britain	95,500
Doñana	Spain	75,500
Ecrins	France	108,000
Gennargentu	Italy	100,000
German–Luxembourg	Luxembourg	72,500
Gran Paradiso	Italy	73,000
Hardangervidda	Norway	343,000
Hohe Tauem	Austria	250,000
Lake District	Great Britain	228,000
Lake Sevan	Russia	150,000

Location	Country	Area (hectares)
Lemmenjoki	Finland	280,000
Neidere Tauern	Austria	75,000
Nordfriesisches Wattenmeer	Germany	285,000
North York Moors	Great Britain	138,000
Ovre Anarjakka	Norway	139,000
Ovre Dividal	Norway	74,000
Padjelanta	Sweden	198,500
Peak District	Great Britain	142,000
Sarek	Sweden	197,000
Snowdonia	Great Britain	219,000
Stelvio	Italy	137,000
Stora Sjofallet	Sweden	128,000
Sumava	Czechoslovakia	167,000
Tatransky	Czechoslovakia	77,000
Urho Kekkonen	Finland	253,000
Yorkshire Dales	Great Britain	176,000
NORTH AMERICA		
Armando Bermudez	Dominican Republic	76,500
Badlands	United States	195,000
Banff	Canada	664,000
Big Bend	United States	283,000
Darién	Panama	597,000
Everglades	United States	566,000
Glacier	United States	410,000
Grand Canyon	United States	272,000
Grand Teton	United States	124,000
Great Smoky Mts.	United States	210,000
Gros Morne	Canada	194,000
Guanacaste	Costa Rica	70,500
Hawaii Volcanoes	United States	93,000
Inagua	Bahamas	74,500
Isle Royale	United States	215,500
Jasper	Canada	1,088,000
Katmai	United States	1,656,000
Kluane	Canada	2,201,500

Location	Country	Area (hectares)
Kootenay	Canada	138,000
Mount Ranier	United States	97,500
Nahanni	Canada	476,500
Northeast Greenland	Denmark	70,000,000
Northern Ellesmere Island	Canada	3,950,000
Northern Yukon	Canada	1,017,000
Olympic	United States	363,000
Pacific Rim	Canada	147,000
Polar Bear Pass	Canada	81,000
Riding Mountain	Canada	297,500
Rocky Mountains	United States	106,500
Sequoia	United States	163,000
Sierra Maestra	Cuba	500,000
Wrangell-St Elias	United States	5,339,500
Yellowstone	United States	898,500
Yoho	Canada	131,000
Yosemite	United States	308,000
SOUTH AMERICA		
Amazonia	Brazil	1,000,000
Banados del Este	Brazil	200,000
Canaima	Venezuela	3,000,000
Chiribiquete	Colombia	1,000,000
Defensores del Chaco	Paraguay	780,000
Galapagos	Ecuador	691,000
Iguaçu	Argentina	1,100,000
Isiboro Sécure	Bolivia	1,000,000
La Neblina	Venezuela	1,360,000
Los Alerces	Argentina	263,000
Los Glaciares	Argentina	446,000
Manu	Peru	1,533,000
Nahuel Huapi	Argentina	428,000
Sierra Nevada de Santa Marta	Colombia	383,000
Vincente Pérez Rosales	Chile	220,000

Krebs, C.J. *Ecology*, Third Edition (Harper and Row, New York, 1985)

Leakey, R. *The Making of Mankind* (Michael Joseph, 1981)

Lean, G. and Hinrichsen, D. *Atlas of the Environment* (Earthscan Publications, 1994)

Leggett, J. (ed) *Global Warming: The Greenpeace Report* (Oxford University Press, Oxford and New York, 1990)

Markham, A. *A Brief History of Pollution* (Earthscan Publications, 1992)

McCracken Peck, R. *Land of the Eagle* (BBC Books, London, 1990)

Mellanby K. *Farming and Wildlife* (William Collins and Sons, 1981)

Mellanby, K. *Waste and Pollution* (HarperCollins, London , Glasgow, Sydney, Auckland, Toronto, Johannesburg, 1992)

Moore, D.M. *Green Planet: The Story of Plant Life on Earth* (Cambridge University Press, 1982)

Moore, D.M. (ed) *The Collins Encyclopedia of Animal Ecology* (William Collins and Sons, 1986)

Myers, N. *The Primary Source: Tropical Forests and Our Future* (Norton, New York, 1984)

Porritt, J. (ed) *Save the Earth* (Dorling Kindersley, London, 1992)

Putman, R.J. and Wratten, S.D. *Principles of Ecology* (Croom Helm, London, 1984)

Rackham, O. *The History of the Countryside* (J.M. Dent and Sons, London, Melbourne, 1986)

Rees, R. (ed) *The Mitchell Beazley Family Encyclopedia of Nature* (Mitchell Beazley International, London, 1992)

Rose, C. *The Dirty Man of Europe* (Simon and Schuster, London, Sydney, New York, Singapore and Toronto, 1990)

Silcock, L. (ed) *The Rainforests – A Celebration* (Barrie and Jenkins, London, 1989)

Simon, J.L. and Kahn, H. (ed) *The Resourceful Earth* (Basil Blackwell, Oxford, 1984)

Thompson, G. and Coldrey, J. *The Pond* (William Collins and Sons, 1985)

Thurman, H.V. *Introductory Oceanography* (Merrill, Columbus, Ohio, 1988)

Whitfield, P. and Pope, J. *Why do the Seasons Change?* (Hamish Hamilton, London, 1987)

Zohary, D. *Domestication of Plants in the Old World: The Origin and Spread of Cultivated Plants in West Asia, Europe and the Nile Valley* (Clarendon Press, Oxford, 1988)

In the mid-1900s, animals and plants were becoming extinct at the rate of several species a day. Most extinctions result from the activities of humankind, usually either habitat destruction or overhunting. Elephants, for example, have long been hunted for their ivory although, since a world ban on trade in ivory was agreed in 1989 by member nations of the Convention on International Trade in Endangered Species (founded in 1973), the fate of the African elephant seems more promising. There are several ways to protect endangered species. Plants may be grown in botanical gardens, and animals bred in zoos to be released into the wild. But some animals, such as the giant panda, are very difficult to breed in captivity.

ENDANGERED PLANTS

Latin name	Common name	Location
Abies nebropdensis	Sicilian fir	Sicily
Agave arizonica	Arizona agave	Arizona (USA)
Ancistrocactus tobuschii	Bandera County ancistrocactus	Texas (USA)
Argyroxiphium kauense	Kauai silversword	Hawaii (USA)
Ariocarpus agavoides	Neogomesia cactus	Mexico
Artemisia granatensis	—	Spain
Asimina rugelii	North American pawpaw	Florida (USA)
Begonia socotrana	Saiberbher	Socotra (South Yemen)
Berberidopsis corallina	Michay rojo	Chile
Betula uber	Virginia round-leaf birch	Virginia (USA)
Camellia granthamiana	Grantham's camilla	Hong Kong
Caralluma distincta	—	East Africa
Catharanthus coriaceus	—	Madagascar
Centaurea junoniana	—	Canary Islands (Spain)
Cephalanthera cucullata	Hooded helleborine orchid	Crete (Greece)
Clianthus puniceus	Lobster claw	New Zealand
Cordeauxia edulis	Yeheb nut	Ethiopia
Crinum mauritianum	Mauritian crinum lily	Mauritius
Cupressus dupreziana	Tarout cypress	Algeria
Cycadales	Cycads – 14 of the 168 existing species of this order are endangered worldwide	
Daphne rodriguezzi	—	Balearic Islands (Spain)
Darwinia carnea	Mogumber bell	Australia
Dicliptera dodsonii	—	Ecuador
Dirachma socotrana	Dirachma	Socotra (South Yemen)
Dorstenia albertorum	Caiapia	Brazil
Drypetes caustica	Bois de prune blanc	Mauritius
Echium pininana	Pininiana	Canary Islands (Spain)
Erica jasminifolia	Jasmine-flowered heath	Southern Africa
Eucalyptus argophioia	White gum	Australia
Euphorbia cameronii	—	Somalia
Gladiolus aurea	Golden gladiolus	Southern Africa
Gunnera hamiltonia	Hamilton's gunnera	Stewart Island (New Zealand)
Herrania balaensis	Cacao de monte	Ecuador
Hesperomannia lydgatei	Kauai hesperomannia	Hawaii (USA)
Isoetes louisianensis	Louisiana quillwort	Louisiana (USA)
Jatropha costaricensis	Costa Rican jatropha	Costa Rica
Kennedia macrophylla	Augusta kennedia	Southwest Australia
Marattia salicina	Horseshoe fern	Lord Howe Island (Australia)
Marojeja darianii	Big-leaf palm	Madagascar
Medemia argun	Dalla	Egypt, Sudan
Microcycas calcocoma	Palma corcho	Cuba
Mimosa lanuginosa	Snow mimosa	Brazil
Myosotidium hortensia	Chatham Island's forget-me-not	Chatham Island (New Zealand)
Neowawraea phyllanthoides	—	Hawaii (USA)
Palmaceae	Palms – 91 species of the family are endangered	
Paphiopedilum druryi	Drury's slipper orchid	India
Pediocactus knowltonii	Knowlton cactus	USA (New Mexico)
Pelargonium cotyledonsis	Old father live forever	St Helena
Persea theobromifolia	Caoba	Ecuador
Punica protopunica	Socotran pomegranate	Socotra (South Yemen)
Rhizanthella gardneri	Underground orchid	Southwest Australia
Saintpaulia ionantha	African violet	Tanzania
Sarracenia oreophila	Tennessee green pitcher plant	Tennessee (USA)
Senecio hadrosomus	Flor de mayo	Canary Islands
Serianthes nelsonii	Hayun lago	Guam, Western Pacific
Tecomanthe speciosa	—	New Zealand
Trochetiopsis erythroxylon	St Helena redwood	St Helena
Ulmus wallichiana	Wallich's elm	Afghanistan, India, Nepal, Pakistan
Vateria seycellarum	Bois de fer	Seychelles

ENDANGERED MAMMALS

Latin name	Common name	Location
Addax nasomaculatus	Addax	Sahara, Sahel
Allocebus trichotis	Hairy-eared dwarf lemur	Madagascar
Balaena mysticetus	Bowhead whale	Arctic, Atlantic and
Balaenoptera musculus	Blue whale	Pacific Oceans
Bettongia penicillata	Brush-tailed bettong	Australia
Bos grunniens	Wild yak	Central Asia
Bos sauveli	Kouprey	Indochina
Brachyteles arachnoides	Woolly spider monkey	Brazil
Bradypus torquatus	Maned sloth	Brazil
Bubalus bubalis	Wild Asiatic water buffalo	India, Nepal
Bubalus depressicornis	Lowland anoa	Sulawesi(Indonesia)
Bubalus mindorensis	Tamaraw	Philippines
Bubalus quarlesi	Mountain anoa	Sulawesi(Indonesia)
Bunolagus monticularis	Riverine rabbit	Southern Africa
Callithrix aurita	Buffy-tufted-ear marmoset	Brazil
Callithrix flaviceps	Buffy-headed marmoset	Brazil
Canis rufus	Red wolf	Texas, Louisiana (USA)
Canis simensis	Simien fox	Ethiopia
Capra walie	Walia ibex	Ethiopia
Capricornis sumatraensis	Sumatran serow	Sumatra
Caprolagus hispidus	Hispid hare	Himalayas
Capromys angelcabrerai	Cabrera's hutia	Cuba
Capromys auritus	Large-eared hutia	Cuba
Capromys nanus	Dwarf hutia	Cuba
Capromys snafelipensis	Little earth hutia	Cuba
Cephalophus jentinki	Jentink's duiker	Ivory Coast, Liberia
Ceratotherium simum	White rhinoceros	Eastern and southern Africa

Latin name	Common name	Location
Cercocebus galeritus	Tana River mangabey	Kenya
Cervus alfredii	Visayan spotted deer	Philippines
Cervus duvauceli	Swamp deer	India, Nepal
Cervus nippon grassianus	Shansi sika	Japan
Chrysopalax trevelyani	Giant golden mole	Southern Africa
Circopithecus erythrogaster	White throated guenon	Nigeria, Benin
Circopithecus erythrotis	Russet-eared guenon	Nigeria
Circopithecus preussi	Prueuss's guenon	Nigeria, Benin
Coleura seychellensis	Seychelles sheath-tailed bat	Seychelles
Colobus badius	Red colobus	Eastern, central and western Africa
Colobus kirki	Kirk's colobus	Zanzibar (Tanzania)
Colobus satanus	Black colobus	Western Africa
Dama mesopotamica	Persian fallow deer	Iraq
Daubentonia madagascariensis	Aye-aye	Madagascar
Dicerorhinus sumatrensis	Sumatran rhinoceros	Southeast Asia
Diceros bicornis	Black rhinoceros	East and South Africa
Didodomys gravipes	San Quentin kangaroo rat	Baja California (Mexico)
Didodomys nitratoides	Fresno kangaroo rat	California (USA)
Dobsonia exoleta chapmani	Chapman's fruit bat*	Philippines
Elephas maximus	Asian elephant	Asia
Equus africanus	African wild ass	Northwest Africa
Equus grevyi	Grevy's zebra	Ethiopia, Kenya
Equus przewalskii	Przewalski's horse*	China, Mongolia
Eubalaena glacialis	Northern right whale	North Atlantic, North Pacific
Felis concolor coryi	Florida cougar	Florida (USA)
Felis iriomotensis	Iriomote cat	Japan
Felis margarita scheffeli	Pakistan sand cat	Pakistan
Felis pardina	Pardel lynx	Portugal
Gazella cuvieri	Cuvier's gazelle	Northwest Africa
Hapalemur simus	Broad-nosed gentle lemur	Madagascar
Hemitragus jayakari	Arabian tahr	Oman, UAE
Hippocamelus bisculcus	South Andean huemul	Andes
Hylobates klossi	Kloss's gibbon	Mentawai Island (Indonesia)
Hylobates moloch	Javan gibbon	Java (Indonesia)
Indri indri	Indris	Madagascar
Kerivoula africana	Tanzanian woolly bat*	Tanzania
Lagothrix flavicauda	Yellow-tailed woolly monkey	Ecuador, Peru
Lasiorhinus krefftii	Northern hairy-nosed wombat	Australia
Leontopithecus chrysomelas	Golden-headed lion tamarin	Brazil
Leontopithecus chrysopygus	Golden-rumped lion tamarin	Brazil
Leontopithecus rosalia	Golden lion tamarin	Brazil
Lepus flavigularis	Tehuantepec hare	Mexico
Liberiictis kuhni	Liberian mongoose	Liberia, Ivory Coast
Lipotes vexillifer	Yangtze River dolphin	China
Macaca pagensis	Mentawai macaque	Mentawai Island (Indonesia)
Macaca silenus	Lion-tailed macaque	Southern India
Macrotis lagotis	Greater bilby	Northern Australia
Mandrillus leucophaeus	Drill	Western Africa
Marmota vancouverensis	Vancouver Island marmot	Canada
Megaptera novaeangliae	Humpback whale	Most oceans

Latin name	Common name	Location
Microtus californicus	Amargosa vole	California (USA)
Monachus monachus	Mediterranean monk seal	Mediterranean
Monachus schauinslandi	Hawaiian monk seal	Hawaii (USA)
Monachus tropicalis	Caribbean monk seal	Caribbean
Muntiacus feaei	Fea's muntjac	Burma, Thailand
Mustela nigripes	Black-footed ferret	USA
Myotis grisescens	Gray bat	Southeast Asia
Myrmecobius fasciatus	Numbat	Australia
Neotragus moschatus	Zanzibar suni	Zanzibar (Tanzania)
Nyctimene rabori	Philippines tube-nosed fruit bat	Philippines
Onychogalea fraenata	Brindled nailtail wallaby (Australia)	Queensland
Oryx dammah	Scimitar-horned oryx	Sahara, Sahel
Oryx leucoryx	Arabian oryx	Middle East
Panthera uncia	Snow leopard	Asia
Panthera tigris	Tiger	Asia
Pan trogolodytes	West African chimpanzee	Western Africa
Pentalagus furnessi	Amami rabbit	Japan
Phalanger lullulae	Woodlark Island cuscus	Papua New Guinea
Phoca hispida saimensis	Saimaa seal	Finland
Phyllonycteris major	Puerto Rican flower bat *	Puerto Rico
Platanista minor	Indus River dolphin	Pakistan
Pongo pygmaeus	Orang utan	Borneo, Sumatra
Presbytis comata	Javan leaf monkey	Java (Indonesia)
Presbytis francoisi	Tonkin leaf monkey	Indochina, southwest China
Presbytis johni	Nilgiri leaf monkey	Southern India
Presbytis potenziani	Mentawai leaf monkey	Mentawai Island (Indonesia)
Procyon gloveralleni	Barbados raccoon	Barbados
Pteropus livingstonii	Comoro black flying fox	Comoros
Pteropus rodricensis	Rodrigues flying fox	Mauritius
Pteropus samoensis	Samoan flying fox	Samoa
Pygathrix avunculus	Tonkin snub-nosed monkey	Vietnam
Pygathrix brelichi	Yunnan snub-nosed monkey	China
Pygathrix nemaeus	Ghizhou snub-nosed monkey	Vietnam, Laos
Pygathrix nigripes	Red-shanked douc monkey	Vietnam, Laos
Rhinoceros sondaicus	Javan rhinoceros	Java (Indonesia)
Rhinoceros unicornis	Great Indian rhinoceros	India, Nepal
Romerolagus diazi	Volcano rabbit	Mexico
Saguinus oedipus	Cotton-top tamarin	Colombia
Saimri oerstedi	Central American squirrel monkey	Costa Rica
Scuirus niger	Fox squirrel delmarva	Maryland (USA)
Selenarctos thibetanus	Baluchistan bear	Iran, Pakistan, Cameroon
Sigmodon arizonae plenus	Colorado River cotton rat	California (USA)
Simias concolor	Pig-tailed langur	Mentawai Island (Indonesia)
Solenodon cubanus	Cuban solenodon	Cuba
Solenodon paradoxus	Haitian solenodon	Hispaniola (Haiti)
Sus salvanius	Pygmy hog	India, Nepal
Tapirus indicus	Malayan tapir	Southeast Asia
Tarsius syrichta	Philippine tarsier	Philippines
Trachypithecus leucocephalus	White-headed black leaf monkey	Southern China
Tragelaphus buxtoni	Mountain nyala	Ethiopia
Zalopus californianus	Japanese sea lion	Japan, Korea

* denotes species which may already be extinct

ACKNOWLEDGMENTS

Picture credits

1 SPL/Claude Nuridsany and Marie Perennon **2–3** PEP/ Kurt Amsler **2–3 inset** SPL/US Department of Energy **4l** SPL/Dr Jeremy Burgess **4r** SPL/NASA **6bl** OSF/ London Scientific Films Limited **6br** OSF/ Okapia/I Gerlach **6–7** Life File/NASA **7t** AOL **7b** FLPA/M Walker **48–9** BCL **50** SPL/NASA **51** OSF/Martyn Colbeck **52–3** Life File/NASA **54** FLPA/Dr D P Wilson **54–5** RHPL **55** SPL/NASA **56** BCL/Jane Burton **57t** OSF/Michael Leach **57b** Ecoscene/W Lawler **58** SPL/Roger Ressmeyer, Starlight **58–9** SPL/Peter Menzel **59** SPL/Peter Menzel **60c** Z **60bl** Ecoscene/ Sally Morgan **60br** Ecoscene **61tl** NHPA/B & C Alexander **61tr** BCL/Stephen J Kraseman **61bl** Ecoscene/Brown **61br** NHPA/David Woodfall **62t** OSF/John Paling **62b** Z/Damm **63** PEP/Kurt Amsler **66t** OSF/Animals Animals/Breck P Kent **66b** NHPA/Stephen Dalton **67** Ecoscene/Cooper **68–9** Ecoscene **71** OSF/Avril Ramage **72t** SPL/ Dr Jeremy Burgess **72b** Biofotos/Heather Angel **74–5** OSF/Michael Fogden **75bl** BCL/M Timothy O'Keefe **75br** Biofotos/Heather Angel **78–9** RHPL **80c** SPL/Claude Nuridsany & Marie Perennon **80b** IP/ Victoria Ivleva **80–1** NHPA/Stephen Dalton **82t** SPL/ Jerry Mason **82c** NHPA/J & M Bain **84t** SPL/NRSC Limited **84b** BCL/Mark N Boulton **84–5** Ecoscene/ Chinch Gryniewicz **86** SPL/Adam Hart–Davis **86–7** Z **87** Ecoscene/Sally Morgan **88** OSF/Belinda Wright **90** SPL/US Department of Energy **90–1** SPL/Roger Ressmeyer, Starlight **91** FSP/Patrick Landmann **92–3** BCL/John Shaw **94t** SPL/Dr R Legeckis **94b** BCL/Michel Roggo **94–5** BCL/Rinie Van Meurs **95** NHPA/Anthony Bannister **96** BCL/Dr Eckart Pott **96–7** OSF/Photo Researchers Inc/Pat & Tom Leeson **97c** FLPA/F Hartmann **97b** Biofotos/Soames Summerhays **98–9** BCL/Konrad Wothe **99bl** NHPA/ David Woodfall **99br** Ecoscene/Brown **100** NHPA/ Stephen Dalton **100–1** BCL/Dr M P Kahl **102–3** OSF/ Partridge Films Limited/Carol Farneti **103t** Ecoscene/ Brown **103b** Ecoscene/Sally Morgan **104c** BCL/Jack Stein Grove **104b** OSF/Howard Hall **105** OSF/Howard Hall **106–7** OSF/Okapia/K G Vock **108bl** OSF/Wendy Shattil & Bob Rozinski **108–9** Z/UWS **110** OSF/ London Scientific Films Limited **110–1** OSF/Animals Animals/Marty Stouffer **112–3** OSF/Edward Robinson **113** OSF/Tim Shepherd **115b** OSF/John McCammon **115t** RHPL/Margaret Collier **116–7** Allsport UK Limited **117** HL/Jeremy A Horner **118–9** Biofotos/ Heather Angel **120** SPL/Simon Fraser **120–1** Ecoscene/ Michael Cuthbert **121** SPL/Peter Menzel **122** FLPA/M Walker **122–3** Explorer/Jean Paul Nacivet **123** NHPA/ Michael Leach **124** NHPA/Hellio Van Ingen **124–5** Biofotos/Heather Angel **125t** AOL **125b** Ecoscene/Eric Schaffer **126** SPL/David Scharf **126–7** Art Directors/Chuck O'Rear **127** OSF/Dinodia/ Isaac Kehimkar **128–9** Explorer/Christine Delpal **129t** NHPA/Nigel Dennis **129b** Ecoscene/Hulme **130–1** Ecoscene/Schaffer **132** FLPA/Panda/ M Calandini **132–3** Photo Researchers/Tom McHugh **133** NHPA/Silvestris/Karl Heinz Jorgens **134** Ecoscene/ Sally Morgan **134–5** Nature Photographers/S C Bisserot FRPS **136t** Panos Pictures/Rod Johnson **136b** NHPA/ Martin Wendler **136–7** RHPL **137tr** SP/Mark Edwards **137br** SP/Mark Edwards **138t** NHPA/Steve Robinson **138bl** AOL **138br** WWF Switzerland **139t** OSF/ Okapia/I Gerlach **139l** OSF/Andrew Plumtre **139b** OSF/Kathie Atkinson **140l** HL/Choco **140–1** Ardea/Wardere Weisser **141tr** Ardea/Jean Paul Ferrero **141br** PEP/Jonathan Scott **142tl** SPL/Will McIntyre **142bl** Biofotos/Heather Angel **142–3** OSF/ David B Fleetham **143l** SPL/Jerry Mason **143r** Countrywide Photographic Library/Derek G Widdicombe

Abbreviations

b = bottom, **t** = top
l = left, **c** = center, **r** = right

AOL   Andromeda Oxford Limited, Abingdon, UK
BCL   Bruce Coleman Limited, Middlesex, UK
FLPA   Frank Lane Picture Agency, Suffolk, UK
FSP   Frank Spooner Pictures, London, UK
HL   Hutchison Library, London, UK
IP   Impact Photos, London, UK
NHPA   Natural History Photographic Agency,
         Sussex, UK
OSF   Oxford Scientific Films Limited, Oxford, UK
PEP   Planet Earth Pictures, London, UK
RHPL   Robert Harding Picture Library, London, UK
SP   Still Pictures, London, UK
SPL   Science Photo Library, London, UK
Z   Zefa Picture Library, London, UK

Artists

Mike Badrocke, Robert and Rhoda Burns, John Davis, Bill Donohoe, Sandra Doyle, Ron Hayward, Trevor Hill/Vennor Art, Joshua Associates, Frank Kennard, Ruth Lindsey, Colin Rose, David Russell, Leslie D. Smith, Ed Stuart, Peter Visscher

Editorial assistance

Jill Bailey, Peter Lafferty, Ray Loughlin, Lin Thomas, Claire Turner

Index

Ann Barrett

Origination by

HBM Print Ltd, Singapore;
ASA Litho, UK